Tool Management Strategies

By
Martin Plute

HANSER GARDNER
PUBLICATIONS

Hanser Gardner Publications
Cincinnati

Plute, Martin, 1946—
Tool management strategies / by Martin Plute.
p. cm
Includes index.
ISBN 1-56990-247-X (hc.)
1. Machine-tools. 2. Production management. I. Title.
TJ1185.P527 1998
670.42'3--dc21 98-27327
CIP

While the advice and information in *Tool Management Strategies* are believed to be true, accurate and reliable, neither the author or the publisher can accept any legal responsibility for any errors, omissions, or damages that may arise out of the use of this information. The author and publisher make no warranty of any kind, expressed or implied, with regard to the material contained in this work.

Hanser Gardner Publications
6915 Valley Avenue
Cincinnati, OH 45244-3029

1 2 3 4 5 6 01 00 99 98

TABLE OF CONTENTS

PREFACE

Productivity is today's lifeblood of corporate survival. If management does not have quick, accurate information upon which to base decisions, guesswork will ensure the equivalent of corporate suicide in the marketplace.

The tool crib, which is usually taken for granted, is one of the last areas in manufacturing to be considered for automated control. The activity of tooling has always presented a major stumbling block for computer automation, and any actual inventory control of the tool crib has consisted of manual record keeping that provided almost no accountability for a tool's whereabouts on the shop floor. The actual activity and use of tooling is confirmed by the act of issuing or returning it through the tool crib. Using this system, no amount of planning, scheduling, or vendor activity can predict with certainty the usage of any tool, and any accurate reporting about tooling usage must be based on the history generated through these manual records.

When you can account for the whereabouts of your tools, you have the basics of a tool management system. If you don't know where they are, you can't report, account for, charge, plan, order, or control their use.

The term Tool Management has been used to describe any activity that includes tools as a resource that directly or indirectly influences the outcome of a specified product or process in manufacturing. Since manufacturing cannot be accomplished without tools, any implementation of control that is directed toward modifying or influencing the use of tools has an immediate effect on the shop floor and ultimately on every department in a manufacturing facility. Following this to its ultimate outcome, tooling affects the financial stability of any company because its use, efficient or not, contributes to the cost of manufacture. If this is true, then how can any manufacturing manager, plant manager, general manager, or chairman of the board ignore the impact of this perishable asset in their company? This is a good question; but for a majority of manufacturers, tooling is an unknown entity or something that is delegated to the responsibility of "someone else." This results in the most serious offense: when tooling is not managed by anyone, it is not controlled by anyone, and it is rarely held to any accountability to anything. Even though tooling is the lifeblood of a manufacturing company's profitability, it is often ignored because it is stored in areas that are understaffed, considered not important to the company's ongoing development, not inundated with high technology, and unkempt in any accountable order.

High technology tooling that cannot properly be utilized by today's machines and processes is being introduced to the market almost daily. To efficiently take advantage of the manufacturing technology that is ushering us into the 21st century will require that tooling, in all its uses and applications, be responsibly managed and accounted for. Remaining competitive in a world market that is being defined by those companies wishing to still be in business into the next century will require speed of manufacture, quality of manufacture, and the ability to respond to the market's needs. Today, tooling information is the least known element about the manufacture of the product. Tomorrow it will be one of the most important pieces of information if manufacturing is to adapt the manufacturing process to meet new design needs, produce to a quality that is acceptable, and still market the product at a price that the consumer is willing to pay.

To control this resource, a computerized tool management system — or several computerized systems working together in the different aspects of manufacturing and company management activity that can actually bring the tool management solutions together — must be employed. Why several systems and not just one system? Because no one system has been developed (and probably will not be in the next twenty years) that is able to make the decisions and provide the information that is required of the ideal tool management system. How can I make this statement? Consider these reasons: 1) the diversity of industries cannot be accommodated by one system or philosophy, 2) every manufacturing company believes itself so unique that no one other than themselves can define a system so strategic to their needs, and 3) the lack of information about tooling in use today is so complete that the discovery of tool usage information will completely confuse the first implementation teams. Their confusion will invite second guessing of their original tool management philosophy, preventing implementation for that time period. In another words, we will keep moving the goalposts back so we will never develop the ideal system.

The linking of existing systems today will allow a productive definition of the future ideal system. There are systems in place that involve the use of tooling today. The expansion of tooling's role in these systems and the capture of tool information from them can be implemented in existing tool management products currently available. What systems exist today that can help us coordinate tooling information and empower that information for decision making? There are CAD systems, CAM systems, MRP systems, scheduling systems, inventory systems, work in process systems, shop floor data collection systems, purchasing systems, maintenance systems, planning systems, and production control systems; and there are others that you could probably think of that I have omitted.

Today, as the cost of tooling rises and the need for tooling becomes more immediate, the value and availability of tools within the tool crib are becoming important factors. Tooling can represent a substantial part of a company's yearly capital investment. Control of this expenditure reaps not only cost savings in the purchase of tools, but it impacts all production areas with potential cost reductions and improved efficiency.

The most important facts that are being revealed by companies that have installed a tool control system are the reports of savings (greater than 20%) in new tooling purchases, and the substantial savings in production time, labor, and process planning. The benefits of a tool control system are so noticeable that the most common comment/question is:

Why did we wait so long to put tool control in the shop?

Probably the most important fact is that tool management system customers having been reporting savings greater than 20% in new tooling purchases. These users also report savings that have provided them with 6-month and less return on investment paybacks. If these facts sound hard to believe, it is probably because you do not have the information you need regarding your current expenditures or tool usage. One of the goals of this book is to point out areas you may investigate to obtain that information.

I know that you may not read this entire book, nor will you conduct complete enough surveys to adequately justify the savings that you feel are appropriate. The general concepts presented herein should be enough, however, to convince you that tool control can contribute a very large savings in both dollar expenditures and production time. The most important message you can send through your organization is that tooling is a major cost savings area and that a company goal should be to get tools under some basic accounting control.

Try to stay away from the "blue sky" syndrome and the "let's interface it with every department that has a computer" traps. Stick with the basics and get tools under control first, then see if the rest of the company is ready to use the information that the tool management control system can provide. This is the biggest stumbling block that you face in the justification process and implementation of a tool management system. The problems that you have today exist because you don't have your tools under control. When you finally implement your control system, these problems will vanish, and you will have a system in place to help you prevent and solve tomorrow's problems.

Good luck!
Martin Plute

But on the whole man is a tool-using animal. Weak in himself of small stature, he stands on a basis, at most for the flattest soled, of some half square foot, insecure enough, has to straddle out his legs, lest the very wind supplant him. Feeblest of the bipeds! 3 Quintels are a crushing load for him; the steer of the meadow tosses him aloft like a waste rag. Nevertheless, he can use tools, can devise tools: with these the granite mountain melts into light dust before him; he kneads glowing iron as if it were soft paste; seas are his smooth highways; winds and fire his unwearying steeds. Nowhere do you find him without tools; without tools he is nothing, with tools he is all.

Tuefelsdröckh

The benefits of a tool control system are so noticeable that the most common comment made at companies that implement tool management is —
"*Why did we wait so long before doing this?*"

The most important message you can send through your organization is —
"*Tooling is a major area for cost savings, and getting tools under control should be an important company goal.*"

Section I

The Basics Of Tool Management

— Chapter 1 —

TOOL CONTROL BASICS

INTRODUCTION

The term *tool control system* has been used to describe the function of storing (in an organized fashion), dispensing, and retrieving, upon request, from the storage area those items called tools. Tool Management adds the requirement of gathering data concerning the transaction of the tools in and out of the tool crib. The act of analyzing, reporting, and using the information gathered for decision making is the application of Tool Management.

In the early 1980's, Tool Management was defined as consisting of tool design, usage control, and tool inventory control. While tool design and usage control team up to help identify standards and eliminate or slow down the proliferation of similar or duplicate tooling, tool inventory control has been related almost exclusively to the traditional central tool crib storage and issue control function.

To emphasize the importance of tools and tooling, consider this statement: a *drill press cannot drill a hole without a bit*. And usually a drill jig, vise, or fixture is required to hold the workpiece while the hole is being drilled. Generally the tools that actually make contact with the workpiece are all too often taken for granted. Only when shortages or bottlenecks occur during the manufacturing process do we begin wondering why we hadn't paid more attention to this *perishable asset* earlier in the planning aspect of production scheduling. Yet many of us do not comprehend the wide variety of tools that can create emergency conditions during production.

The following list of tools is not all inclusive and will change according to the manufacturing process requiring tools, but it does provide an overview of the complexity of tool management.

- Hand tools for manual operation, such as assembly operations, final adjustments, installing fixturing, testing, and other manual functions.
- Foundry patterns, cams, molds, templates, progressive stamping dies, powder metal dies and their components.
- Workholding tools, vises, chucks, collets, fixtures, components to fixtures, angle plates, jigs, nuts, and bolts.
- Perishable tools, i.e., end mills, drills, inserts, grinding wheels, taps, saws, reamers.
- Designed tools, i.e., EDM electrodes, form tools, broaches, dies, punches.
- Hard tooling (durables), i.e., tool holders, boring bars, fixturing , adapters, dies, mandrels, centers, tool blocks, and their components.
- Inspection tools, gages, testers and test equipment, surface plates, parallels, checking fixtures, measuring equipment.
- Miscellaneous items: handy wipes, shop rags, gloves, tapping oil, safety glasses, shoes, uniforms, marking tools, light bulbs, and toilet paper.

In conventional (traditional) machining, the tool management function is generally considered under the control of the operator or production engineer of the product. The tools are usually specified and purchased by these individuals, and their use to manufacture the part is decided by one or two machinists. The tools are ground and fabricated according to the machinist and his needs. This was the general methodology and tool control from the early 1800's up to the mid 1960's. Some manufacturing facilities still use this methodology.

The next stage in tool control began as many other departments became important to accomplishing and completing the transformation of a raw material into a finished product or service. Tool design became more formalized and specialized. Production engineering, process planning, and purchasing began influencing the availability and type of tooling that was available for use. Gantt charts and production schedules began appearing, along with burden rates, cycle time requirements, piece part rates, and cost analysis. Tooling was usually looked upon as in the crib, and if there was a production bottleneck it was because the tool crib didn't have any more tools. This problem still exists in many manufacturing facilities today.

When Computer Aided Manufacturing (CAM) and CNC (Computer Numerically Controlled) machines began to appear, the problems became more critical because of the faster machining capabilities of these machines. Scheduling and production control had a higher price tag now, but if everything went as planned, the payback was very attractive. However, tooling was still a predominate contributor to production delays. One inescapable detail that was misunderstood for much too long was that timely information was not available about tooling, and consequently tooling was still considered a minor function under control of the tool crib. The solution for tool availability was simple, but not recognized until 1980.

As CAM gave way to CIM (Computer Integrated Manufacturing), control of the machine and its tooling moved away from the operator on the shop floor and was taken over by the engineering and production management. As computers became part of manufacturing, the learning curve of how to use them, and what to use them for, introduced more confusion and misdirection. The promised improvements and simplifications that were hoped for from this new technology are still being sought today even though we are now beginning to harness the power of information gathering and the manipulation of that which we have acquired.

A major factor in the unfulfilled application of these super brains in manufacturing was the ability to properly integrate and communicate the necessary information to those people who needed it during the manufacturing process. Information is the keystone for control, and receipt of information at the proper place and time is the implementation of control. A myriad of information processing software has been evolving since the 1960's. In all these software and computer based application products for manufacturing, the term "tool" has usually defined the machine itself and *not* the cutting element in the spindle or any other mechanism that actually shapes the workpiece. In general, any provision for tool information in these application packages addresses machining and production specifications, or cost, vendor, and other planning information. The ability to schedule the tool's availability was ignored except for some theoretical projections using raw inventory status as it applies to infinite material requirements logic. What is slowly being recognized is that tool management —knowing where tooling is employed at any given time — is a major ingredient in the successful manufacturing process. The simple action of recording the issue and return activity of tools sets the stage for integrating information management with other manufacturing application systems. The availability of tooling is affected by tool design, tool engineering, production control, CAD, CAM, purchasing, process planning, product design, and manufacturing engineering. The most important factor is the actual utilization of a tool, as this has the final affect on its availability. This is the factor that most tool management philosophies overlook. This simple step of Tool Management sets the stage for Tool Resource Planning, Tool Standardization, Tool Information Control, and the integration of these strategies.

WHY CONSIDER TOOL MANAGEMENT?

Accuracy in data collection about tool activity is necessary if we are to provide any reliable means of tool control. But tooling presents unique problems to the process of collecting data.

There are many reasons for controlling tooling, but none is as dramatic as the reduction of manufacturing costs. The impact of reducing this cost through tool management is one that many are slow to recognize. However, a *Metalworking Production Survey* study conducted in the United States in 1987 reveals some significant facts concerning the percentage of tool costs in manufacturing. The three following statements are excerpts from that study.

- In the manufacture of high volume products, vehicles, and white goods, tool costs range from 2 to 4 percent of the manufacturing cost.
- In the manufacture of medium volume products or medium batch part runs, valves, compressors, etc., tool costs range from 6 to 8 percent of the manufacturing cost.
- In high technology and low volume parts or product manufacture, aircraft, sub-contract, etc., tool costs range from 8 to 12 percent of the manufacturing cost.

Other findings include the following.

European studies that predate the U.S. study suggest that the cost of tooling in manufacture of products is closer to 33 percent, and the remaining costs are split between labor and machine tools.

Manufacturers do not have a good record for managing tool inventories. Tooling, primarily cutting tools, represents close to a $4 billion/year market. This is about half of the market for machine tools.

The cost of tooling-up machines can represent 20 to 33 percent of the machine tool's cost, with machining centers being the most costly.

The most obvious reason to consider tool management is inventory control. This area provides the most natural and identifiable starting point for tool management. The immediate results of reducing inventory and new tool purchases provides the basis for understanding the potential for tool control elsewhere in the company.

Additionally, accurate inventory control allows for the discovery of obsolete tooling (typically, up to 50 percent of the inventory value) and the identification of lost or missing items, which can represent from 30 to 60 percent of the inventory value. Almost all tool crib managers who have applied computerized inventory packages to the tool crib have stated savings of 20 percent or more saving of durable tooling and 40 percent of perishable or consumable tools.

Process planning benefits from tool management include the ability to monitor tool usage and performance. This ability assists manufacturing when determining the optimal tooling requirements for future work. Additionally, monitoring tool performance allows tooling improvement relationships to be rationalized by including feeds and speeds, tool life, and tool change times in the selection process.

Finally, tool management assists the reduction of machine downtime. Ensuring tool availability reduces waiting time for tools during production and setup.

— Chapter 2 —

AUTOMATION AND THE TOOL CRIB

Tool management's primary objective is to provide the right tools together with the right information at the right place at the right time.

As advanced manufacturing techniques and technology appear in the factory, management responds with automated systems to monitor, control, and report the information flow from and to the equipment. The sophistication of these new management systems requires technicians and manufacturing managers to develop computer skills never before required in the manufacture of products.

The rapid growth and implementation of new technology on the shop floor is making its impact throughout the world. The tool crib is one of the few places within the shop or manufacturing plant that often seems to function as it had in the past, although there are some improvements being made such as increased use of modular storage cabinets, automated storage carousels, inductive chips, tool dispensing machines, automated parts storage, bar code readers, magnetic stripe readers, and information retrieval devices. However, the overall operation and management of the tool crib, and the tooling stored within it, remain relatively unchanged.

The activity of a tool crib and its inventory have, in the past, presented major stumbling blocks to automating its control. The record keeping for a tool crib traditionally has been a manual task, which can be extremely difficult to manage and operate. These existing methods, coupled with the necessity to work with personnel who are unfamiliar with automated control systems, creates many human interface problems. The actual use and life process of a tool creates accounting, storage and tracking problems not easily reconciled by existing management (partially resulting from management personnel's lack of knowledge of shop floor tool activity) and the existing system software.

However, if an automated system were available to provide immediate information on a tool's inventory status, it would enable a manufacturing facility to increase production, reduce inventory costs and machine downtime, and give shop floor management an important tool for monitoring the activity and relationship of tooling to the actual production of parts. It would also provide a substantial savings that would merit immediate attention.

BASIC TOOL CRIB FUNCTION

The basic role of the tool crib is to store the tools and supply them, when required, to the machines on the shop floor. How this is done varies from shop to shop. Generally, when a tool is issued to the shop floor, some sort of ticket or slip of paper is used to record the transaction. Ideally, an inventory adjustment is made to record the issue of that tool from the storage location to its assignment to the machine, job, and the user requesting it. As long as tools are readily available, the shop floor and production schedules run at the predicted pace, and few problems or bottlenecks are experienced. Only when tools are not available do managers start thinking of ways to bring the inventory of a tool crib under some means of management control.

If you are using a manual card system, you simply write a new inventory count on the card to keep track of the on-hand quantity. Of course, this can not be done reliably for every tool issued to the shop floor, nor is it easy to quickly identify upcoming shortages.

If you are using a KANBAN type of inventory control, an ordering procedure will have been established, and when a certain inventory level is reached, the card is read, faxed, copied, or otherwise recorded and sent to a supplier or your purchasing department, and the restock process is started for those tools. This is an improvement over the card system, but it is still a manual sorting and reporting process that is not actively updating the status of the inventory. Also. there is no verification of the whereabouts of the tooling once it leaves or where it has been when it returns to the tool crib.

Another popular solution is to build a simple database of the tool inventory on a personal computer using one of the popular database building and management products like dBase®, Foxbase®, Access®, and a multitude of others. Once this is designed, a copy is put in the tool crib and the attendant is instructed to update the counts of the tools in the database as they are issued and received. This is a step in the right direction, but generates more problems than it solves, Unless, the only thing you wish to solve is *how many of each tool item is in the tool crib at a given moment.* But to handle emergencies, you will also want to set minimum reorder points for each tool, etc., until the simple system becomes a patchwork of requirements that make it a complex system that is still unable to accurately project needs for each tool. If basic inventory is all you are after there are several low cost inventory products available commercially for as low as $10. You may already have an inventory system in use on your mainframe or network system. But, of course, we know traditional inventory control systems for raw materials, finished goods, or in-process parts do not cover a tool crib's operational needs for cutting tools, fixtures, gages, and supplies, otherwise you would not be attending seminars or reading this book.

The basic requirements for tool crib control dictate that not only the *quantity* but also the *whereabouts* of the tooling be known. The most natural and accurate place to obtain this information is at the **issue point**. This requirement means that the tool crib attendant must input this information to a tool control or accounting system. Depending on who or what department wants to know about the disposition and availability of tooling determines how the information is requested and what the crib attendant will be required to input. In most home-grown attempts to build a tool crib control system, this information is highly tailored to the department designing the system for the company. In reality, the only information required to build an effective tool control system is that which the crib attendant normally inputs. He is probably writing this information on a slip of paper — namely, the tool number, how many issued, who requested it, for what job and which machine.

The crib is the issue point for most tool transactions. The tools may be sent and issued from other areas, i.e., preset areas, other cribs, cell areas, FMS stations, assembly areas, etc., but, they should and do originate from a central storage or distribution/receiving point. Recording the actual quantity and user of the tools from this issue point establishes an accuracy benchmark. You cannot assume that since you have been told by a Process Planning or MRP system that five tools are allotted to this job or operation, that five tools and only five (not three or seven) were actually issued. What if two broke during setup or for some other reason? Then seven tools would have been issued. This is a fact, and if you know these facts it could make a difference in the cost accounting of that job for that machining operation, for the evaluation of the setup technique, and any future tool use planning for that job. However, you now have two fewer tools in the actual physical inventory. Some job in the future that needs this tool will be delayed, and it is your future production bottleneck.

If you look at an average company and follow the life and death of a tool and identify the systems that track that tool's life and death, you can begin to form a strategy to use this information and implement a method to continue gathering this information. A tool's life begins in tool design as a required element in the manufacture of a product starting with the product's design. There are many CIM systems that have requirements for this knowledge; and through their use of this information they contribute to the database of available information concerning this tool. Process planning and production control systems also require similar information during their planning session and contribute to the information about the need for the tool. Their information requirements include how many tools will be needed, and where and when they will be needed for the efficient and economic production of the part to a schedule that was probably predetermined by the material requirement's planning and finite scheduling systems. Purchasing needs to order the tooling or request it to be built, which means the tool must be identified, then stored, and finally issued for use on the shop floor. It will probably return and will require inspection, refurbishment, or repair. The quantity will be monitored, if only to confirm the available quantity in the drawer before the next production run, and it must be tracked to its usage point and discarded when its useful life has been consumed. If the company used its shop floor work-in-process and inventory control packages to monitor this activity, it would then be able to evaluate the tool's contribution to the productivity of that product.

We all know of existing computerized systems in our manufacturing facility that capture or are capable of capturing a good majority of tooling activity information that has been identified in the previous paragraph. Logically then, it should be a simple task to download, consolidate, communicate, or just print out some of the facts we have acquired about the tool. The reality is that this information is usually incomplete, poorly stored, conflicting, and most often discarded. Tooling is one of the most overlooked, mismanaged, and ignored assets in manufacturing. It can be said that the typical manufacturing facility lives with the following truths concerning their control of tooling.

- There is no control of tooling.
- Upper management is generally ignorant of tooling's importance; in most cases, they don't know what a tool is or where it is stored.
- Production experiences delays because tools are unavailable.
- An overabundance of unneeded tools is available.
- Duplication of tooling exists to excess.
- Each employee's tool box or work station is a complete miniature tool crib.
- The ability to apply JIT philosophy to tooling is incomprehensible.

The common approach to tool management is when Joe Doaks, our tool crib manager who has been in the tool crib since the building was built, looks in the cabinets and says it is time to reorder, so we reorder. If it's not in the tool crib, it must be in the building, and we will search for it no matter how long it takes or how many people look for it. And if we can't find it, St. Anthony will. It is during these crisis occasions that somebody suggests that a tool management system might be something to consider. But as soon as the crisis is over, tool management is no longer on the docket.

GENERAL EXISTING TOOL PROCEDURES

A tool becomes part of the machine shop or manufacturing floor's inventory in a variety of ways. The path that determines the birth of a tool depends on whether it is purchased standard, modified from purchased standard, designed special and made in-house, or purchased made to print by an outside vendor. The following numbered steps illustrate that no matter how tool requirements or design specifications might arise, the procedure to procure it for production is always similar.

1. A need for the tool is identified as what would fulfill the production requirements of the product being manufactured. This may be recognized by a CNC programmer, a machinist, a foreman, process engineer, tool engineer, or the product engineer.
2. A request to purchase or build the tool is issued.
3. The requisition is reviewed and either modified, rejected, or approved. Eventually, an approval of this or some other replacement of the tool is approved.
4. The approved requisition goes to purchasing.
5. A vendor is selected or approved for the tool and a purchase order is issued.
6. The tool is shipped and received, inspected, approved, and delivered to the tool crib or the requester.
7. A tool card is created recording the primary tool description, purchase requirements, storage location, and quantity.

At this point, the tool is considered available for use on the shop floor for the specific functions it was designed or purchased. However, there are no controls in place to monitor the actual disposition of the tool or its performance. It could become lost, dull, broken, or scrapped, and no one would really know for sure. Every so often, or when it is needed again, it may be requisitioned again if the inventory seems low or if a routine check of the available tools in the crib is conducted for reordering purposes.

If this tool is a fixture, tool holder, or collet, it could completely disappear. There is no control whatsoever paid to durable tooling. If it is scrapped or lost, it will not be noticed until the job that requires it is run again.

The simple definition of a tool control problem stems from the previous procedure description. This tool is but one of maybe 4,000 perishable tools, or one of 8,000 fixtures, tool holders, collets, or other durable type tools. If it is a specially designed tool or modified standard, it could be one of 40,000 specials or in-house-defined tools. Now, if the average cost of a perishable was only \$2.00, and a durable tool only \$150.00, and a modified in-house-designed tool only \$90.00, then this inventory alone is only worth \$4,808,000. This, of course, is a fictitious inventory, and if it was real would represent quite an investment for maintaining production capability. But this inventory value is probably lower than most companies in operation today. This inventory must continually be replenished, repaired, resharpened, and calibrated, and it grows as the business grows with new products and production techniques. The problem stems from the simple fact that no control is exercised with any competent reporting or support over this perishable renewable asset of the company.

The consensus of most foremen and machinists who interact or depend on their tooling to be available from the tool crib is that they can depend on the tool crib to supply their needs 70% of the time. Are the tool crib personnel the reason for lost tools, misplaced tools, tool shortages, dull tools, etc., or is it the fault of inadequate control systems provided by upper management to keep adequate inventory available. Simply reordering is not the solution for all the tool shortages. Tracking of the tool's activity and use is the solution to controlling its availability.

The implementation of a tool management system will be difficult, because the solution is not just inventory counting. The existing situation in the tool crib has been going on for years in some companies, and the operation of the tool crib has not changed since the doors opened. The bad habits established in the tool crib and on the shop floor as to the tool crib's function are hard to break. A tool management system usually requires a computer at some point, and computers can only do what they have been programmed to do. Even the most flexible programmed systems have rules. If these rules are not followed, the results will be flawed or possibly nonexistent.

BASIC ERRORS TO AVOID

There are a few bad habits that are hard to break, but which must be identified and addressed before any tool control system can be implemented.

- Tool numbers are not assigned to the tools. Even though you may use the manufacturer's number for tool holders, collets, drills, end mills, etc., you must have a tool number for each distinct tool item.

- No records exist for the tool's vendor, cost, order quantity, or location. A simple card system could assist in reconciling this and the tool number problem.
- No records exist for scrapped tools. Without monitoring the frequency or existence of scrapped tools, tool shortages can go totally unnoticed. Any recovery method of tools for scrap value is nonexistent.
- On-hand inventory value is not available. Value of durable and perishable tools is unknown, usually requiring an inventory to be taken before any budgeting for future needs can be accomplished.
- Identification of obsolete tools cannot be easily accomplished. Often obsolete tools are included in requisitions and their stock levels continue to increase. If they are issued, they may continue to be restocked by the crib manager and cause manufacturing problems because an obsolete tool was issued for the job.

Tools are stocked in several locations in the crib and on the shop floor. This totally destroys the reorder process. Multiple locations or point-of-use storage techniques promote overstocking and confusion of tool reporting, accountability, and user responsibility. If a tool is not available at one location or at the tool crib, then a search is conducted of all possible other locations. This also makes it difficult for any new crib attendants to learn the storage and operating procedures of the tool crib.

Multiple location storage promotes piecemeal ordering. This is the most popular way to overstock — and overpay — for tooling. It can cause you to pay higher purchasing costs both in paperwork and in per-piece pricing of tools.

If tool crib confidence is low on the shop floor, hoarding will occur. Each machinist has his own personal crib of tools that are critical to him. This includes gages, fixturing, and holders, as well as tools. It also guarantees that the tools are probably dull, out of tolerance, or in need of repair. This condition produces the kind of scrap you do not want from the shop floor.

These problems are very typical in shops where no tool control exists — even traditional manual card systems cannot escape these problems. Simply by investigating your current system, and interviewing shop foremen and operators, you will uncover these problems. If you try to improve the crib's control through a manual record keeping system, these problems will still plague your crib. Even some mainframe inventory control systems cannot alleviate these problems. Why? Because manual systems and some mainframe systems are too labor intensive. In order to function correctly, the inventory must be adjusted at the time the tools are issued, and (this is the part that disrupts most card and simple computer inventory systems) when the tool is returned from the shop floor.

The solution lies in the accuracy of the method chosen to record the tool crib and the tool inventory daily activities. Without accuracy to the chosen solution there will be no credibility. Credibility is one of the primary reasons for implementing control in the tool crib.

Manual systems have difficulty identifying reorder points, but mainframe inventory systems can assist if the stock levels are maintained because they can shuffle through the inventory figures faster than the crib attendant can shuffle through the stack of inventory cards. However, neither system can quickly identify obsolete or overstocked conditions unless steps have been purposely programmed identifying the activity condition. Neither manual or mainframe systems are adept with recording the tool's location status. Once the tool leaves the tool crib, it is usually difficult to quickly identify where it is on the shop floor.

Manual systems do not provide management with the current visibility of tool availability or usage. This information is important to current and future production scheduling or planning. Tool performance, usage, and availability are important byproducts of tool inventory control.

Machine tools cannot produce the required dimensions or product without the correct tools. Each tool required also depends on the tools to hold it in the tool holder. Thus, the loss or unavailability of any tool, tool assembly, or component of the assembly can cause undue production delays. Production delays idle machine tools, machinists, and setup bottlenecks for other manufacturing processes. These delays all add up to preventable costs — if you know what tools are available, they can be provided to the process when needed. Overstocking the tool crib does not provide an economical or guaranteed solution to these problems. Only the control of tooling that is provided through an automated system and the rules that it is designed by can alleviate and reduce tooling related bottlenecks and delays.

SUMMARY

The key words necessary to maintain control of the tool crib are **inventory** and **management**. These functions are most often viewed from the office and scheduling management areas and not on the shop floor or in production management areas. To bring a traditionally manual function under control, such as the issue of tooling, one must view this activity from the user's point of view. Generally, the first mistake made is to handle the tool crib as an inventory control problem. In traditional inventory control, parts or items are stocked and issued, generally never to be seen again. This is not true in a typical tool crib. *Tools are returned.* When returned, they are generally not in a condition that allows them to be readily replaced in their normal storage location. They may require sharpening, inspection, or repair, or they may be broken and unrepairable and must be scrapped. This deviation from traditional inventory logic causes record keeping problems at the crib. To manage the information about the tens of thousands of different tools that could be stored within the tool crib, accurate records of each tool must be updated and readily available. A flow chart of typical tool activity in a shop is shown in Figure 2.1.

To accurately monitor tool availability, a tool management system should be capable of recording the activity, as well as the inventory quantity of a tool as it is used on the shop floor.

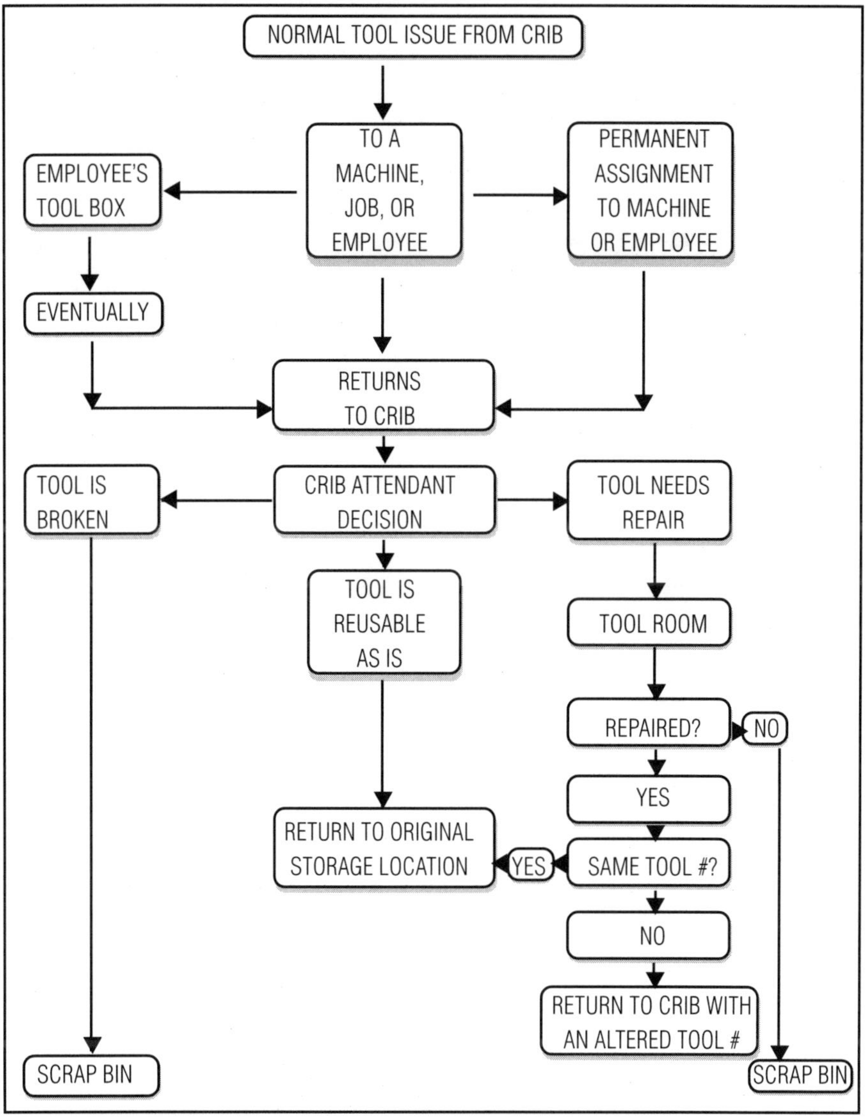

Figure 2.1—Typical tool activity.

— Chapter 3 —

IDENTIFYING THE REQUIREMENTS

Once the problems of your current tool-related operating procedures are identified, they can be used as the criteria for developing or selecting the tool management system required. Based on the previous discussion scenario, a basic tool control system would need to maintain an accurate inventory of all tools and their location in the crib and on the shop floor. This feature alone would eliminate many of the current problems, such as tool shortages and lost tools. It is important to observe that not all tool related problems can be solved by simple tool inventory. To produce the benefits that tool management can render, a system must include the activity of tools. This includes tracking, planning, reporting, organizing, ordering, and other activities that are required by the various personnel that control your tooling requirements.

The many and varied definitions of tool management will be restated for your specific endeavor of tool management analysis. The primary definition usually suffices; "*Provide the right tool at the right time at the right place.*" This should include every tool in the process, purchased, designed, and modified for use on the shop floor or in the production of piece parts. It should also include interface ability, operational input or, at the very least, information pooled from the following departmental areas: tool design, tool crib, purchasing, process/production, and planning and manufacturing engineering. In some instances, you may also consider including the quality control, gage, and metrology areas.

The first consideration or requirement will be the most important step in building the fundamentals that will support the additional requirements of a tool management system. Tool inventory control is the most important basic requirement because it sets the control functions for inventory accuracy. The reason many tools are not available when they are needed for production is because the inventory activity is not recorded, thus the inventory level is not adjusted. The production activity can completely deplete a stock level before the next reorder activity is attempted. A computer system that monitors the issues of tools would minimize this problem by requesting purchase for tools as their inventory level goes below an established minimum stock level.

Inventory adjustment can provide a means for controlling scrapped tools. Normally when a tool is scrapped, the operator or foreman just throws it away and requests another from the tool crib. This process continues until the inventory is depleted. By requiring the documentation of tools when scrapped, and the reissue of additional tools through the tool management system, the tool shortage can be identified quickly, reordered in a timely manner, and a record of the scrap condition can be analyzed in the future. This future analysis can help predict and identify production trends, tool performance expectations, and machine characteristics.

In most cases the control of inventory adjustments is from an input terminal in the tool crib to the computer. The rules of operation of this terminal usually designate that the crib attendant and other personnel directly related to the storage, restocking, issue, and receipt transactions of the tools are the only personnel directly involved with the recording of the tool transactions. The number of people involved in direct input of the transactions should be kept to a minimum in order to reduce the chance of errors entering the computer and therefore increase the accuracy of the inventory. This indicates that security at the crib level is in order and that training of all personnel involved with data or transaction input must be familiar with the computer tool management system and the rules associated with the various transactions.

In order for these steps to be properly implemented, a tool identification and an organized storage philosophy must be in place. In another words, a tool number and description must be provided for each tool to be stored, monitored, and tracked. Also, a means of storing the tools and an identification of where the tool can be found in its storage location must be provided. In addition, min/max levels, cost, tool type, and other basic data need to be created and stored as they relate to each tool.

If the inventory adjustments are being captured during the issue or restocking operation, then the ability to additionally capture the tool's work assignment and its location of use can also be input at the same time. Again, an input terminal at the point where the transaction is occurring can conveniently transmit this information to the computer database. In other words, we can track the travel and use of the tool from the point of issue and return. This additional information enables us to find tools on the shop floor when sufficient quantities in the tool crib are unavailable. When properly implemented, you can locate tools in use at machine tools for specific machining operations, in the possession of employees, or in repair. Without the ability to track all tools and their activity, the tool crib attendant does not know if the tool is even in existence any longer. He may suspect that it is still around, or may even know that it is because of his familiarity with the job, machine, or people using it, but it is probable that no one else in the company would have this information. Tool tracking allows everyone with access to the computer transaction database to retrieve tool usage information pertinent to their job responsibility. When you do not know where tools are, the usual result is duplicate orders, unnecessary reordering, or replacement of the tools. Thus, overstocking and unnecessary tool expenditures become the rule rather than the exception. Another cost is the time involved in searching for the tool. You may know it is in the shop, and you may think you know where it is, so there is a strong tendency to just go out to the shop floor and search for it. The time wasted adds up quickly in indirect labor costs: tool presetters waiting, machine operators waiting, foremen searching, and production bottlenecks forming and becoming real. All this cost can be eliminated by simply tracking the tool's activity at the time of issue or return.

Tool tracking is one of the elements that separates tool management philosophy from the stock and standard inventory philosophy. The need to track tools differs from that of general inventory control of raw materials and parts. Tool tracking requires a procedural commitment from the tool crib personnel, the shop floor personnel, and management. The procedure is to simply record where the tool is issued for each transaction. This simple action prevents hoarding, reduces lost tooling, increases inventory accuracy, and provides the means of controlling that encourages the return of tools. The benefits of tracking tool issues are so obvious that the procedure becomes second nature to all crib, shop, and management personnel.

As controlling individual tool issues becomes common, it also becomes apparent that usually more than one tool is issued to a machine for a particular part machining operation. If all these individual tools are grouped together and issued together, many advantages can be realized at both the crib and machine tool. This process of grouping tools is generally called *kitting*.

Tool kitting can provide a package of tools necessary for the job with the minimum of issue time, and at the same time assure that the correct tools are issued. The implementation of kitting at the tool crib and on the shop floor provides management with an in-process quality control mechanism that reduces scrap and increases production rates. If management maintains control over the building and definition of the tool kit for each machining operation, then management can ensure control over the tools that are issued from the crib to the machine for that machining operation.

A typical CNC machine could average fifteen or more tool assemblies (in some cases, many more), where each tool assembly could be considered a kit and all the assemblies could be controlled by including them as a component list of tools for the machine. This process is called "nesting kits" or "kits included within a kit." Since an average job can require many different tools, assemblies, fixtures, and gages, it is cumbersome and impractical to issue tools individually. The grouping of tools into kits in advance allows the crib attendant to quickly and with minimum data entry issue complete tooling setup packages through designated tool kit or tool package numbers. The key to the kit package number is coordination between the CNC programmers, manufacturing engineering, production engineering, quality control, and tool design to use common numbers and standardized tooling.

One aspect of computerized kitting is that the tools do not have to be dedicated to the tool kit or package, and they do not have to be separated from the crib's inventory and stored together in a separate tool kit bin. In the old days, all the tools for a setup were stored in a bin together. The end result was dull tools, damaged tools, and fixture or tool shortages. Of course, some habits are hard to break, and many companies will build up tool kits, especially tool holder assemblies, and leave them assembled. This can reduce the availability of critical components for other jobs. Usually, it's the lack of visibility that prevents the reuse of critical tool assembly components for use on other tool assemblies and machines. This visibility can be improved through computerized tool kit definition and the tracking of the kit's components (in use and in storage), and especially the tracking of kit components within the storage area for tool assembly and preset. Computerized tool kitting and tool kit tracking can allow many different kit philosophies to exist: breaking tool kits down for storage and reuse, leaving tool kits and their components assembled in the machine, storing assemblies at the machine, and storing assembled tool assemblies in preset or the tool crib.

Computerized kitting helps reduce the impact of a dedicated tool kitting philosophy by providing visibility of the dedicated tooling and the individual components. This information assists in reducing overstocking conditions of tools in kits, reduces redundancies that can occur through dedicated kits, and prevents shortages through tracking tools to their dedicated kit storage locations. However, computerized kitting will not reduce the additional storage space required and multiple locations that will result because of dedicated tool kits.

A way of reducing the multitude of kits that can arise from the many different parts and machining operations is to use a standardized tool kit program. Standardized tool kits are a collection of tool assemblies that are common to many machining operations or parts and are commonly used, stored, or maintained for machines and jobs. For example, to tap a 1/4-20 screw thread on a particular machine for any machined piece part, the common tools that are used for the tap drill kit and tap tool holder assembly could be defined. This set of tools for the 1/4-20 thread could then be recalled for any job that required a 1/4-20 tapped hole. Standardizing common machining operation kits and tool assemblies will reduce machine setup, tool issue time, tool presetting time, tool selection decisions, and planning.

One of the obvious benefits of establishing the aforementioned features and desired procedures is that the computer can allow organization of the data files that manage this information into formats that can provide timely information to management and the shop floor. This capability is commonly referred to as tool management reporting. Without a means of assembling reports of certain tool activity or tool groupings, it would be difficult to make improvements in efficiency, reduce unnecessary inventory, reduce tool operating costs, purchase tools, and monitor tool performance. The ability to define a set of standard reports that can be called upon at any time by any group of interested personnel is a distinct advantage of a formal standard computerized tool management system. Additionally, ad hoc or on demand reporting is also an advantage of computerized tool management. In fact, once the data files are established, and as long as they are updated through (maintained) the tooling transactions at the tool crib, the information contained in them should be readily available to other computer systems and manufacturing control software programs, thus making your tool management system and the activity of tooling a strategic information resource to manufacturing and other information control systems that may be implemented within the company. This strategic information status of tool management is what makes it a political management system. Knowledge is power, and the knowledge of how tooling is used reflects on all the decisions that are made before the actual use of the tool and its resulting successful or unsuccessful use.

The activity of a tool may be mundane to some, but tracking the travels of a tool can supply valuable information about the progress and potential bottlenecks concerning production. Evaluation of tool performance, a machine's operating condition, an operator's ability, and even the accuracy of the process plan can be made from monitoring the tools in use during the production process.

Generally, when a tool control system is considered for the first time, the tendency is to design the system from the top down, because the top needs to manage the information and the information must be shared among several departments. However, the activity information has to be gathered at the bottom to ensure accuracy of the information. At the beginning of all projects, everyone is striving to see the total picture and they want to ensure that their area of expertise will look good in the picture. Figure 3.1 lists features that are usually implemented later in the life of a tool control sys-

Ease of Use
- Data entry
- Flexibility
- Crib attendants
- Managers
- Systems support

Classification
- Perishables
- Durables
- Gages
- Assemblies
- Fixtures
- Returnable
- Nonreturnable
- Auxiliary search code

Reporting
- Dedicated
- Customized
- User definable

Cross References
- Tool cost by job
- Machines
- Jobs
- Users
- Information files
- Vendor files
- Alias (AKA)
- Easy to define
- Quick access
- Easy search

Tracking
- Machine
- Job
- User
- Rework areas
- Subcontract
- Inspection

Kit Control
- Infinite nesting
- Ease of definition
- Ease of issue/return
- Where used kit/tool
- Preset data
- Availability
- Component tracking
- Assembly tracking

Communications
- ASCII files
- Host computers/applications
- with CAPP/MRP/CAM systems
- Purchasing system/vendors
- Inductive chips
- Bar code devices
- AS/RS and tool dispensers

Audit Trail
- ASCII file
- Custom reportable

Reserve and Transfer
- Assign issue conditions
- Transfer machine–machine
- Issue from crib
- Transfer to rework

Figure 3.1.—Tool control key features.

tem but are presented here to provide a look at the bigger picture of tool control. Remember, there still must be a mechanism for capturing the basic information about tool activity and defining its primary database needs. Now that you have seen a list of desirable features, in Chapter 4 we will examine why implementing them may present a few stumbling blocks.

— Chapter 4 —

TOOL REQUIREMENTS PLANNING AND TOOL MANAGEMENT

Tool Requirements Planning (TRP) is usually viewed as the next step to pursue after and sometimes before a tool management system is implemented. Tool Requirements Planning is not a new concept. Its idea is closely related to the inventory control and bill of materials concepts fostered by Material Requirements Planning (MRP) systems. The realization of TRP came during the 1970's when CNC and FMS equipment began to be established in manufacturing. The proliferation of tooling that was needed on the machine to complete the machining process was exceeding the machine's capability.

Tool inventory control and tracking help provide location and order control for the tooling needs, but the overwhelming proliferation of tools indicated that control was needed elsewhere in the manufacturing process. TRP became another part of the definition of what is Tool Management. An MRP system for tooling was needed. Information about what tools were used on what parts and machines, and when they were used, became an information priority. Bills of tools (kits), like bills of materials in MRP systems, became a critical component. Tool inventory control and tracking had benefited from kit control and could now allow product design, production, and manufacturing planning to benefit from this information. The proliferation of tooling had now taken on another dimension apart from the primary goal of reducing inventory — it now had the secondary goal of reducing the number of tools needed to produce the piece part. This secondary goal would be as dramatic as the original goal in reducing inventory.

THE BIRTH OF TOOL RESOURCE PLANNING

One of the first steps is to identify what are standard tools for manufacturing and which are specials. Next, a definition of these standards must be applied by all departments involved in the evolution of the product: Product Design through Tool Design. Another benefit of standardization of tooling for production would be the lowering of tooling cost. More standards means less specials. Even if specials are necessary, the rationalization of the need for specials would reduce their proliferation and associated costs to build them.

The first step in tool management, Tool Inventory Control, is very important in providing the basic information for TRP. The tool inventory can be grouped by tool classification. By this grouping, similar tools can be pointed out regardless of their original numbering scheme (see Section III for more information). The classification grouping allows standard tools to be realized from special order tools; and so lesser known, duplicate, and infrequently used tools are easier to spot, reclassify, or omit. Identification of unnecessary tools helps reduce tool inventory proliferation while giving Product Design and Tool Design a good reading of the tooling that exists within the current inventory.

There have been many papers written on the standardization of tooling needs. Most revolve around the need for standardizing common machining and design concepts; for example, consolidating screw sizes and types to reduce the number of clearance holes, tap drills, and taps, or using common clearance cuts for cutoffs, grooves, slots, and other noncontact geometries. You might want to think of it as Tool Rationalization Planning (another TRP), just so we can maintain three-letter-acronym confusion.

THE REBIRTH OF TOOL REQUIREMENTS PLANNING

Tool Management's visibility is providing a means of revisiting TRP, and it is enjoying a rebirth. Today it is a relatively fresh concept and it is less often considered a component of importance during the manufacturing process plan. Thus, many manufacturing plans have machines, materials, and labor properly scheduled, but the job is held up or delayed because the right tools were not at the right place at the right time. Just In Time (JIT) concepts and plans will fall apart if Just In Time Tools (JITT) are not part of that plan. Tooling is generally an accepted commonplace item. As long as the money is available, we will overstock our tool cribs to ensure that we do not run into shortages. But if the money runs low or the coordination of tool assembly management falls down, then and only then do tools become an important consideration. When the bottleneck or delay is solved, it's production as normal again — no tool control. TRP becomes a considered resource because its concept relates to the concepts offered by MRP systems.

Basically, Tool Requirements Planning indicates what tooling is required, how many tools will be needed, and when the tooling will be needed. Answering these requirements allows TRP to assist the future tool requirements needs of the manufacturing process. With a tool management system in place, TRP can gather current and previous tool usage information and more accurately advise on future tooling needs. One of the most important ingredients that TRP and tool management systems share is the Bill of Tools or Tool Kit list and requirements, which results in a commonality in tool identification and numbering. Information requirements about tools in the TRP system may be shared with the Tool Management system. However, the TRP system will need additional information to complete its exchange of information with MRP. Most commonly TRP is linked to MRP to provide tooling information that assists the MRP system with decisions concerning the actual production schedule.

TRP is ideally suited to those companies that produce their own products or that already employ an MRP system. The benefits of using tooling in a multipurpose or multimachining schedule are more apparent to dedicated parts manufacturers. In addition, TRP allows for the scheduling of higher cost tooling to be amortized over different parts production thereby lowering the impact of the tool's cost. The TRP function is a critical ingredient to a JIT system. For example, the need to reduce machine setup time can only happen if the tools are available and if they have been scheduled to be available on a time critical need. The management of an FMS often illuminates tooling and setup difficulties if an insufficient quantity of tools exists.

It has been mentioned that TRP requires an interface to MRP to function correctly. TRP also needs feedback from a shop floor tool management system that can report not only the inventory availability of tooling, but the actual issues, returns, scraps, and rework conditions of tooling. One of the important data categories of TRP is that of tool life. Using the shop floor production information, tool life can be managed and predicted for the tooling stored in the TRP systems database.

However, some argument can be made that the final interface should include input to a Finite Scheduling System (FSS). This association with an FSS can offer ongoing updates of tooling requirements based on the changes to the production schedule. This allows rescheduling of tools as the interruption of shop orders, due dates, or other changes occur in the production schedule.

Tool Requirements Planning systems can assist in controlling tool availability in the manufacturing schedule. The goal of a TRP system is to increase manufacturing productivity. Another benefit of TRP is that it provides the tool crib with a window of tooling needs for manufacturing. TRP allows a benchmark for comparing tooling costs and the establishment of tooling budgets. It also allows tool cost to become an ingredient in the cost of manufacture, and it justifies the need for duplicate tooling while minimizing redundant tooling. The usage history captured by a shop floor production or tool management system is enhanced, which allows for the allocation of tools according to a predetermined tool consumption rate or against a scheduled need. Finally, it reduces scheduling conflicts of like or similar tooling.

— Chapter 5 —

CONSIDERATIONS FOR A TOOL DATABASE

FOCUS

Tool Management is a term that has different meanings to many manufacturing professionals. Depending on job function, responsibilities, and general experience in manufacturing, Tool Management will be interpreted in many ways. Tool Design, CNC programmers, Production Planning, Manufacturing Engineering, Purchasing, etc., all have different responsibilities concerning tools. A single tool database that would satisfy the needs for these departments would be monstrous. Before starting to assemble a potentially strategic database, step back and consider the elements of the primary goal — to make tooling visible.

DECISIONS

When deciding what information your tool database needs, everyone naturally puts their needs first. If you are planning to meet the needs of all departments, then you have six or more groups that feel that their needs are most important. Since the information contained in a database can be critical, the techniques used by the tool control software must access the information quickly. A complex record-length database requires complex programming techniques. Most tool control procedures defy the stoic logic required by computer programming and will send many database attempts back to the drawing board.

Meetings to cut the fat from the database will help determine: Who is going to give up what? Why is certain information required on a report? What will be the primary tool number used? There are countless other questions. No matter what is decided about the final contents of the database, someone will not be happy. If the dissatisfied party had a database established before, on paper or in a computer, they will continue to use it and attempt to adapt it for their future needs. Fragmentation, incompleteness of the database organization, and dissension among departments is the usual result. The reason for concern is that each department realizes that this information will become strategic to their needs in the future. A solution could be decentralized, separate databases for the different departments, but with one condition: that they can interrelate with each other when necessary.

ORGANIZING

Each department begins by determining the database requirements that will define their strategic niche for the future. However, one strategic department with very little computer experience (and which may never have seen a computer) is wondering what a tool database is. Ironically, this department holds the key to interfacing with all the other departmental databases. Unfortunately, most companies provide very little technical expertise to this critical area — your tool crib and tool stores. It is the place where the tools are stored, issued, reworked, returned to, and requisitioned from. Later in this section we will focus on the database requirements of the tool crib.

Tool database information should be classified or assigned to subject areas of responsibility in tool management functions. Instead of having one database with huge records containing everything possible about the tool, break the record up into files that deal with the subject areas; for example, a tool identification file that contains primary tool descriptive information, and an activity file that contains activity records. Using this technique allows for expansion, by adding new files to the database in the future for the events you did not foresee today. Other examples of files that could be contained in the database are vendor files, bill of material (kit) files, user files, etc.

POTENTIAL PROBLEMS

One problem that could present a stumbling block to organizing your tool database in this file structure is record access speed. Selection of the record management system and database engine will be important. Sequential database engines won't cut the mustard for this technique. They will be too slow and may have file limits that could hamper future expansion or not allow the current size needs for your tool inventory. Identify key fields in the record of each file. This method will allow you to assign indexes to the key fields and will provide a vehicle for programming access to all records and files. This method will give you a speed advantage over other database record retrieval techniques. Remember, your tool database is not going to be small. You need to use proven techniques that work with large database applications and that are easy to expand later. Investigate Relational Database languages and engines that have experience and demonstrated abilities with real time and multi-user applications. Languages whose roots are PC based or Windows based may be adequate for local single user access, but may have serious data management problems when accessed by multiple users simultaneously.

Plan the information that should be contained in each record for each file within the database. You will still run into size restrictions, but the real size restriction will be external to the computer programs. The size restrictions will be on physical input. For example, tool numbers that are too long can produce keyboard input errors for crib attendants and others accessing the information. If you are using bar codes, look at how long the bar code label will be with your 36-character tool number. When planning report layouts, remember that large fields are cumbersome to position on a report, and an overkill of information in the report can lead to multiline records. These factors can make the report hard to read and usually present more information than necessary for the intent of the report.

DATABASE INTEGRATION

By setting up your database in file (subject areas) responsibility, it will be easier to integrate information from one or several files between application programs or computer systems. It may seem more complex on the surface, but it allows for expansion and interrelationships to be associated to the other department's tool database needs. It will allow importing and transferring information containing kit/tool assemblies, pre-set information, CAD/CAM drawings and programs, tool life information, job cost concerning tooling, tool activity, ordering, etc. The information exchange possibilities keep growing as manufacturing moves into the 21st century. A single record database design could never handle or foresee all of tomorrow's interface and database information needs.

PRIMARY INFORMATION

As we discussed earlier, an organized tool control system begins with the building of a database that contains the necessary information for accurate reporting. The primary requirements are the tool number, tool description, location, quantity, cost, vendor, reorder point, and maximum need. Expansion of this basic information would include considerations for referencing more than one vendor; more than one storage location; classifying tools as perishable, durable, gages, fixtures, dies, returnable, or nonreturnable; and assigning alternate or class identification. An example of a typical tool database record is shown in Figure 5.1.

```
Tool: ADRILL________                       Cost: ___1.65

Description: A (.234D) DRILL 2-5/8 FL 3-7/8 OAL__
  Type (P)erishable (D)urable (G)FI (A)ssembly: P
              (R)eturnable (N)onreturnable: R

Vendor: CLEVELAND TWIST_     Reorder Qty.: 36___
   Min. On Hand Qty: 12___   Max. On Hand Qty: 120__

Primary/New Location: CR14 02_   Qty: 9___
Backup/Used Location: ST12 02_   Qty: 0___

Auxiliary ID: DRIL  Dim 1: .234___   Dim 2: 2.625__

Alias: 710-A123456098765-12
Extended Descr.: STANDARD JOBBERS HIGH SPEED STEEL
```

Figure 5.1—Basic tool database information.

Figure 5.1 displays typical tool database information as it would appear on a computer screen. Usage information is created as the tool is issued and becomes part of the database record. It is usually displayed on other report screens, for example, the EDIT screen or Activity Report, and updated as transactions are recorded at the tool crib window.

The next step is to make provisions for gathering tracking information about the tool's activity and location. In traditional inventory logic, you assign a storage location when you define your basic tool information. As tooling is issued to the shop floor, it is customary to record the machine or employee using the tool. This provides shop floor visibility of the tools. Tooling also has the unusual inventory characteristic of returning to its point of issue (or at least it is desirable that it return). If it does not return, overstocking is occurring without your knowledge and a strong cost savings benefit is lost. Forcing the use of tools through their usable life cycle and recycling tools provides a large cost savings in itself. For these and other reasons, it is usually desirable to know to whom, for what, and where the tooling has been reassigned.

Traditionally, some form of tool checkout slip or tool loan receipt has been used for the tool checkout at the crib window. The Tool Loan Receipt records, in writing and in multiple copies, the employee number, the machine, the work or job order, the date, the quantity, and the tool number. There are all kinds of variations of the tool checkout slip; some contain more information, some less, but they all basically do the same thing — they provide some traceability for locating the tool after it leaves the crib. We all believe that some day we will need that tool and want this information so we can find it. Examples of a typical tool withdrawal transaction and the resultant transaction ticket are shown in Figures 5.2 and 5.3.

Figure 5.2 displays information requested for the issue of a tool to the shop floor. The highlighted (**bold-face type**) information is input by the crib attendant. Machine, job, and user information are for tracking and return accounting.

Machine: [**1045**]__ Job: [**A87650**] User: [**3045**]__

Tool: [**ADRILL**]________

Description: A (.234D) DRILL 2-5/8 FL 3-7/8 OAL

Type: Perishable - Returnable

Quantities available: Primary/New: 9 Backup/Used: 0

Quantities to withdraw: Primary/New: [**2**]____ Backup/Used: 0

Figure 5.2—Issue tools to the shop floor transaction request

A transaction ticket is printed for the crib attendant specifying the storage location of the tool (Figure 5.3). The transaction ticket could be used as a chit and given to the machinist. Notice that a *reminder to reorder* is given on the ticket.

Now that we have the tooling where we want it, it returns. This is where traditional inventory rules start falling apart. When a tool comes back to the crib or preset area, some decisions have to be made by the crib attendant. "Is it returnable to its original storage location?" "Should it be repaired, sharpened, or scrapped?" *These questions must be answered to complete the in-*

```
  |                                                              |
O| WITHDRAW 06AUG98                                             |O
 | MACHINE: 1045    JOB: A87650    USER: 3045                    |
 |                                                               |
O| TOOL   DESCRIPTION                     LOCATION QTY          |O
 |------ --------------------------------- --------- -----       |
 | ADRILL A (.234D) DRILL 2-5/8 FL 3-7/8 OAL CR14 02  2 NEW      |
O| *Onhand qty. is below reorder level - notify supervisor*     |O
 |                                                               |
```

Figure 5.3—Printed withdraw transaction ticket.

Tool: [**ADRILL**]________

Machine: [**1045**]__ Job: [**A87650**] User: [**3045**]__

Description: A (.234D) DRILL 2-5/8 FL 3-7/8 OAL

Type: Perishable - Returnable

Total Qty. Withdrawn: 2 Qty. Being Returned: 2

Quantities returned to: Primary/New: 0 Backup/Used: 0

Qty. to Scrap: [**1**]____ Qty. to Rework: [**1**]_____

Rework Dept: [**DEPT54**]

Figure 5.4—Tool crib window returns transaction ticket.

formation cycle of a tool traveling from its storage location to the shop floor and back to the tool crib. In order to maintain an accurate count of all the tools in inventory, the travels of the tool must be recorded.

A typical tool Return Transaction ticket is shown in Figure 5.4. The condition of tools returned to the crib can be easily recorded, as shown by the example. Only the highlighted information needs to be entered by the crib attendant, and the Machine, Job, and User blanks may be automatically recorded by the Tool Location Control system.

When tools are returned, the transaction ticket (Figure 5.5) provides the crib attendant with a record for the disposition of the tools. When a tool is returned to the crib and reassigned, other questions begin to appear. "Do I put used tools in with new tools?" "When a tool is returned to the crib from a rework or refurbishing operation, is it still the same tool it was when originally issued?" Simple and straightforward rules can be established for answers to these questions that can be easily understood and enforced by the crib personnel. Establishing rules for handling tool transactions is an important step toward identifying the different data requirements needed in your Tool Location Control database.

A typical return from rework and the transaction ticket is shown in Figures 5.6 and 5.7. One problem that takes tool control out of the standard inventory control classification is that repaired tools, when returned to inventory, may have changed from their original specifications. The Return from Rework ticket provides areas for the new tool description to be entered (see Figure 5.6) thereby providing a solution for this problem.

IMPLEMENTATION

After establishing the kind of information needed to define the tool and how to track its whereabouts, the next task is to input the data to the tool control system. The basic tool description and information makes up the primary database which may be extracted from existing databases or directly from the tool crib records.

In order to assure that the location of the tool can be traced, all records of tool activity must come from the point of issue, the tool crib window itself. To maintain accuracy it is essential to capture the information concerning the

```
 |                                                                |
O| RETURN  06AUG98                                                |O
 | MACHINE: 1045   JOB: A87650   USER: 3045                       |
 |                                                                |
O| TOOL  DESCRIPTION                       LOCATION QTY           |O
 |------ --------------------------------- -------- -------       |
 | ADRILL A (.234D) DRILL 2-5/8 FL 3-7/8 OAL         1 SCRAP      |
O| ADRILL A (.234D) DRILL 2-5/8 FL 3-7/8 OAL  DEPT54 1 RWRK       |O
 |                                                                |
```

Figure 5.5—Printed return transaction ticket.

Tool: **[ADRILL]**___ Rework dept: **[DEPT54]**
Description: A (.234D) DRILL 2-5/8 FL 3-7/8 OAL
Type: Perishable - Returnable
Total Quantity in Rework: 1 Qty Returned: 0

Qty primary/New: 0 Qty backup/Used: 0 Qty Scrap: 0

To be returned as New tool: [**ADRILL-RG1**]_____
Description: A (.234D) REGROUND SHORTER ADRILL
Type: Perishable - Returnable

Qty primary/New: 0 Qty backup/Used: [**1**]

Figure 5.6—Return from rework transaction.

```
O|                                                                    |O
 | RETURN FROM REWORK  06AUG98                                        |
 | REWORK DEPT.:  DEPT54                                              |
 |                                                                    |
O| TOOL        DESCRIPTION                  LOCATION  QTY             |O
 |----------- ------------------------------ --------- -------        |
 | ADRILL-RG1  A (.234D) REGRIND SHRTR ADRILL CAB1 D2   1 USED        |
 | ADRILL       WAS THE ORIGINAL TOOL                                 |
O|                                                                    |O
```

Figure 5.7—Printed rework transaction ticket.

activity directly from the issue point, as illustrated by the previous transaction examples.

Since the tool crib attendant issues the tools, he becomes the most important link in recording their activity. The crib attendant has always been the most important person in the management of tool inventory. He has the primary responsibility for inputting and controlling the information that comprises the tool database. His accuracy establishes the reliability of your reporting, reordering, tracking, etc. A good analogy is to consider your tool crib attendant as a bank teller and your tool crib the bank vault. When you go to the bank to withdraw money, the teller has to have your identification and record the transaction. I have never been able to walk into a bank and just ask for fifty dollars without giving something to the teller, but I have gone up to a tool crib window and taken half a dozen inserts without accounting for them. You need to implement the same precautions at your crib that you find in a bank.

Tool: ADRILL Description: A (.234D) DRILL 2-5/8 FL 3-7/8 OAL
Type: Perishable - Returnable

(Tool On Order)

Reorder Point: 12 Unit cost: 1.650
Auxiliary ID: DRIL Dim1: .234000 Dim2: 2.62500

Location		Qty	Value
CR14 02	Primary/New	9	14.85
ST12 02	Second/Used	0	0.00
	Totals	**9**	**14.85**

In-Use At -

Machine	Job	User	Qty	Date	Value
A2377	956-C	9897	1	3MAY98	1.65
B2445	4234-A	PRESET	2	1AUG98	3.30
E8771	655-L	3112	1	27FEB98	1.65
		Totals	**4**		**6.60**

Reserved for

Machine	Job	User	Qty	Date	Value
A7112	877-B	*	1	14JLY98	1.65
C238	656-A	7091	1	19JLY98	1.65
		Totals	**2**		**3.30**

Rework Dept.	Qty	Value
ACMETL	1	1.65
TOOLRM	5	8.25
DEPT54	3	4.95
Totals	**9**	**14.85**

Figure 5.8. Typical tool status report.

Reporting

If all the tool management rules and procedures have been followed, then the status of a tool can be reliably reported upon request. All locations where the tool can be found become common knowledge. The Tool Status Report (Figure 5.8) and other reports are now available and they will be accurate because the information needed was captured at the point of issue — *the tool crib window*.

When you can account for the whereabouts of the tool in the crib, on the shop floor, or in rework, you now have enough information to build a basic tool control system. This information then becomes important to the next progressive step in tool control — tool management reporting. Tool management reporting will lead to tool resource planning and other sophisticated reporting. This building block approach has proven to be a successful way to obtain your future tool management needs. So far, we have only taken the first step, getting the tool crib and the whereabouts of the tools under control.

Remember, if you don't know where the tools are, you can't report, account, charge, plan, order, or control their use.

When you can account for the whereabouts of your tools, you have the basis of a tool management system.

— Chapter 6 —

TOOL LOCATION AND INVENTORY RULES

Now that you have an understanding of the flow of tools in and out of the tool crib, and a procedure for capturing the information, a few basic commonsense rules need to be setup before your tool management system can be put into motion. The following rules will provide the foundation upon which a successful tool management system can be built.

The Ten Fundamental Rules of Successful Tool Management.

1. **Control the access to the crib**. Put a lock on the door. A fence would help.
2. **Organize the storage cabinets**, shelves, or boxes that are used to store the tooling. Give the locations names, tags, or some sequence of identification (numeric, alphabetic or both).
3. Insist that **all tooling issued and returned is recorded** through the system. This ensures a higher degree of accuracy.
4. **Involve the crib personnel from the beginning** of your tool control project. They know more about how the tooling is issued than anyone else and can provide information critical to the building of the tool database.
5. **Train all tool crib personnel**. Provide at least two competent system administrators for overall system control. Ensure that someone

who understands the complete system operation can be reached by all operating shifts.

6. **Establish guidelines for** the **return** of tools to the crib for **rework** consideration, for **scrap**, or return as **new or used** tooling.
7. **Review your purchasing procedures**. Hold a joint meeting with all personnel involved in the requisitioning and purchasing of tooling. With tooling under control, piecemeal ordering should become the exception not the rule.
8. **Establish rules for defining perishable and durable tooling**. A suggested guideline could be as follows.
   Perishable refers to tools that can be consumed by use: i.e., drills, end mills, taps, carbide inserts, etc..
   Durable refers to tools that are generally not consumed by use: i.e., tool holders, collets, micrometers, fixtures, power tools.
   Returnable refers to tools that are expected to be returned to the crib after use: i.e., a perishable/returnable could be an end mill that can be reground and reused; and a durable/returnable could be a set of micrometers or a tool holder that is to be stored in the crib after use.
   Nonreturnable refers to tools that are not required to be returned to the crib after use, or tools that are permanently assigned to a department/machine or employee: i.e., a perishable/nonreturnable could be small diameter drills, gloves; or a durable/nonreturnable could be special tool holders or fixtures assigned to a specific machine.
9. **Establish rules for reordering perishables and durables**. A suggested guideline to control reorder points could be as follows.
   For **Perishable Tools:** When the total quantity of tools in the crib inventory is below the defined minimum quantity, it is time to reorder. Used tools in the crib are included in crib inventory; in-use tools and tools in rework are not included as part of the available inventory.
   For **Durable Tools** (including gages, fixtures, dies and assemblies): When the total on-hand quantity of tools is below the defined minimum quantity, it is time to reorder. Tools located in the crib and on the shop floor are included in the total. The exception is nonreturnable tools, where only the crib quantity is considered.
10. **Agree on a tool numbering system and stick to it**. Establish a group to be responsible for the implementation and maintenance of the numbering procedure.

Once tooling is brought under control, the tool database now becomes ***strategic*** to others. Access to information about tools by other departments will be a ***necessity.***

INPUTS AND INFORMATION CONTROL POINTS

In order to obtain the information required for reports, your tool control system should allow the personnel responsible for tool information to easily record all tool activity. There are two basic groups or levels of responsibility that provide the information about tooling that will build your reports — your crib personnel and your tool supervisory personnel. Their basic input functions will allow you to monitor your tool activity and status through a variety of reports. Suggestions and explanations of reports, from the information available, are explored in Chapter 10 on Tool Control Specifications.

You may begin your tool control system with a simple personal computer system. As you become more familiar with the system and fully confident in the accuracy of the reports, you will begin to explore other uses for the system; for example:

- the control, tracking, and recall of gages and inspection tools,
- the ability to place tools on a reserve or advance need status,
- the ability to track fixtures and dies,
- the tracking and replacement of tool assemblies and magazines, and
- the interface to presetters and FMS control systems.

You will also realize that you will need to expand your tool control system to allow simultaneous input of tool information by the crib window attendant(s), supervisors, and planners, and for access to the reports. This means expanding your PC microcomputer by upgrading to a multiuser system.

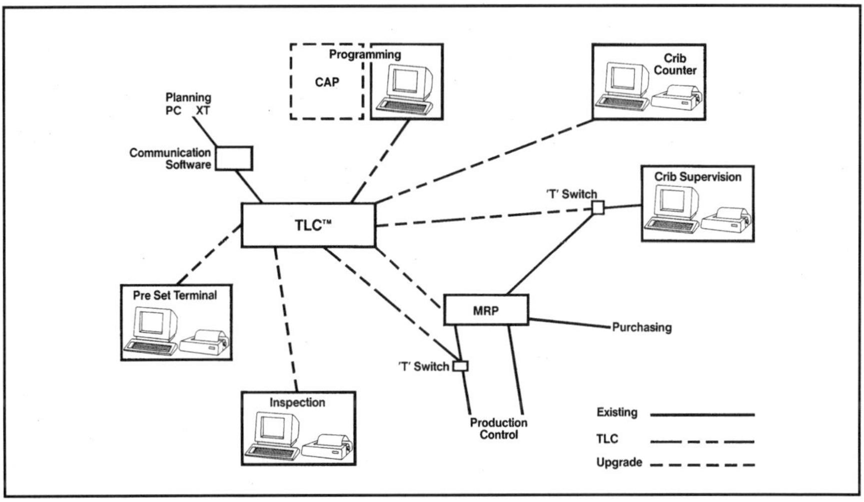

Figure 6.1—Typical tool management systems installation.

A tool control system typically consists of a supervisor's control station and remote terminals at the crib window with other management areas connected to the supervisor's microcomputer. This inevitable future expansion can be easily implemented provided you have built the basic requirements into your tool control system. Figures 6.1 and 6.2 illustrate a typical tool management system and possible variations.

All tool management systems differ from company to company because of intended use, number of users, etc. However, they all typically contain the following similarities: 1) they all have a supervisor terminal and report printer located in the tool crib superintendent's office, and 2) they all have terminals with ticket printers located at the issue and return windows of the tool crib.

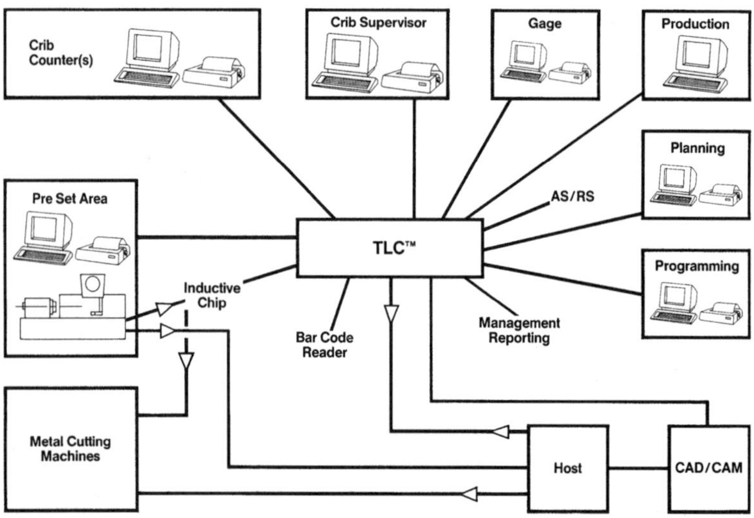

Figure 6.2—Tool management system installation variations.

Common variations to this configuration are: 1) a terminal for process planning or the N/C programming group for tool kit definition and tool availability reporting, 2) a terminal for tool design for the addition of new tools and the management of special tools, 3) a terminal or interface for purchasing to facilitate final approval or confirmation of tool requisitions, and 4) an AS/RS or carousel interface controller and terminals for retrieval and pick list queuing of tools and kits. There may also be requirements for terminals in rework and preset areas. This configuration can be expanded further to include FMS integration as shown by Figure 6.3.

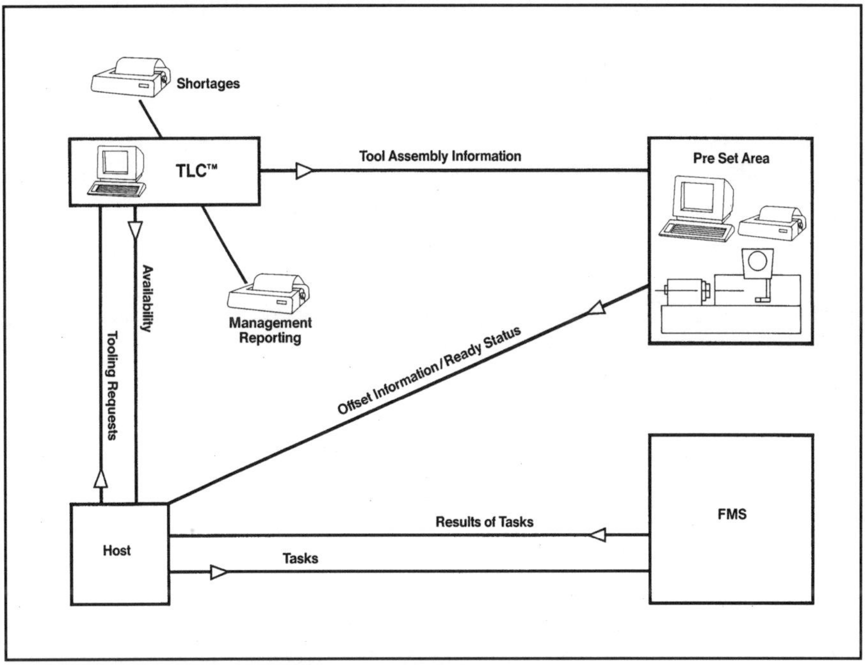

Figure 6.3—Typical FMS and preset station installation.

IMPLEMENTATION OF THE TOOL MANAGEMENT SOFTWARE AND HARDWARE.

The next step is to implement the functional requirements into a formalized software system for use on a computer. To accomplish this step, three basic approaches will present themselves to most companies: 1) develop the software programs in-house, 2) modify an existing software program, or 3) purchase a commercial tool control product. Each of these approaches has its advantages and disadvantages. Each company must evaluate its own particular needs, software support technicians, and manufacturing environment in order to pick the approach that, in their opinion, will provide the greatest benefit. The following discussion is meant to give some insight to these approaches.

Develop the software programs in-house

This is the most time-consuming and expensive of the three approaches. It necessitates extensive amounts of programming man hours (or man-years) and dedicates them to design and implement efforts that have previously been done by other companies. This approach is called, for obvious reasons, *reinventing the wheel.* Time is also spent duplicating concepts that exist in other systems. The primary reason this approach is selected is if your needs are so specific that they cannot be met unless the software is completely tailored to the uniqueness of your company's operating procedures. This approach will give you exactly what you require for tool management abilities.

If you do not have the personnel in house to develop the software, you can write a specification and hire an outside development group to write the software. This is more costly but necessary if your normal business is not writing software systems. An outside service must produce the finished product to your specification and usually will contract at a defined and known end price for the software. Support and changes can be contracted for ongoing development and service. Either way — in-house personnel or outside service group — it is going to cost a great deal to support and write the software from scratch and provide ongoing development and support.

The initial payback period and the implementation date will be longer with this approach. Unless you are writing a system in some simplified software language, the cost for the approach can easily exceed $500,000. This is, of course, a multiuser system running on a respectable computing system. If you are considering something that uses a PC based network with a few files to do inventory control and order tools, you should consider the second approach (below) in this series.

The drawbacks to doing the work in-house are numerous. If you did not select this approach for the reasons given earlier (that your specific tool control needs can not be supplied or satisfied any other way than with tailored software), then any other reason was the wrong one. Let me give you one of the biggest wrong reasons — N.I.H.

The N.I.H. (Not Invented Here) syndrome means that you or your Information Systems manager has the attitude that if it isn't built by this group or implemented by your personnel and tailored for your company specifically, it can't be good enough for your needs. This means that you probably build your own tools, machines, desktop computers, CAD/CAM Systems, etc. Or it means that you are looking for some work to keep the IS department in business or to hire more personnel, or you need a reason to buy more computer power for the IS group. If your business is not building software products, then skip this approach and stick to your business. If your business is building software products, then you already know that you need personnel who understand the intended use and expected result of the tool control system. If your IS personnel do not understand what is expected of the application software, then you are inviting disaster.

Modify an existing software program

This approach should be considered when an existing in-house system already provides several of the mechanisms and features that you feel are primary to your tool control needs. An existing system may have an item master for a bill of materials, and with some redesign, a tool inventory and tool kit list could be a possibility. Another possibility is that the existing system can monitor manufacturing parts, raw materials, and production processes, and modification to these existing systems is necessary because — as they stand now — they almost fit the tool control idiosyncrasies, but not quite. If you have an existing system that can meet your tool control needs by modifying the system, then your tool control needs aren't very demanding. Unless a package is specifically designed to control tooling, it's usually limited in some fashion or can barely supply the basic control necessary to conduct the needs of tool inventory. In general, it is safe to say that the existing systems will assume that tools are available to run the scheduled job or that they can be ordered, or have been designed, etc., but in reality the opposite is true. To modify these systems is to change the premise by which the systems were originally designed to function. The modification of the original premise of the system's design may be detrimental to its original purpose. Remember, the rules for managing tool inventories are not the same as those for managing information and inventory of maintenance items, raw materials, finished goods, or parts in process.

However, if you can modify an existing system, the advantage would be that it will cost less and require less time and man power than developing a system from scratch or contracting a customized system. The in-house personnel are familiar with the existing system and could modify it with less effort than designing a new one would require. Work on the modification can begin (and stop) upon release of the modification specifications (or remodification of specifications). Another advantage is training. Less training is required because of familiarity with the existing system. Additionally, any computer or other system or communications interfaces that are necessary are already in place.

There are a couple of disadvantages to this approach. The changes that need to be made may not be practical when considering the intricacy of modification to the existing database and data files. Depending on the complexity of the existing system, the modifications could affect the operability of the existing system. The sayings "a bird in hand is worth two in the bush" and "if it ain't broke don't fix it" come to mind at this time.

Purchase a commercial tool control product

Purchasing a commercially developed system does not let you off the specification hook. You still need to consider your tool control and management requirements and define a specification list. Because of the number of commercially available systems, your list will help narrow the viable products and give vendors the ability to understand how well you have studied the problem. Hopefully, the selection will be no more difficult than comparing and choosing the one that most closely fits your requirements list. However, since you have not used or implemented a tool management system before, some vendors may disagree with the specific requirements you are asking for. Some of them will disagree because *they* can't meet the need; others will argue that your need is not real and will disappear once you have your tools under control. So, direct evaluation of the available packages is a necessity. Try systems out, visit users, attend seminars, or ask for in-house trial periods.

This approach has many advantages over the other two choices. The biggest are cost and experience. In short, time, money, man power, facility, and resources are not wasted in reinventing the wheel. Implementation is easier and can proceed on a defined schedule, because others have done this before you with the same software. The software is continually being improved by the vendor and you can generally subscribe to some form of maintenance or upgrade policy. The software also provides the benefit of support from the vendor for training, installation, and innovative tool management techniques. Since several companies are using the software, the odds of experiencing programming errors are less and reporting of errors is better. This means that the software will be improved by the use of more and more customers providing more field proven suggestions for improvement than would be possible if you were the only user of the software.

There are also some disadvantages even to this approach. Since this system is dedicated to tool management, it is an island — a stand alone software product. Unless there have been some specific steps to transfer or accept information from its database, interfacing other in-house programs could be difficult if not impossible. The development of interface software could be costly or impractical under current system constraints. Even if the database is transportable to other applications, the applications may not easily accept the information.

UPGRADING AND FUTURE CONSIDERATIONS

If you choose to build your own tool management system, the future aspects are only as good as your ongoing research and practical experience. The new trends that you innovate in your home grown system may in fact be outdated methods to the rest of the industry. If you are modifying an existing system, then your modifications to that system for future needs are only as good as the flexibility of the existing package's original design goals and end use function.

Future enhancements to tool management systems are going to be discovered by innovative users of systems through continually developing and incorporating new techniques for improving tool management and control. This capability alone is a good reason to consider the purchase of commercially available tool management software products. The convenience and cost savings associated with the purchase of commercially available stand-alone software outweighs many times over the fewer benefits that are realized by the other two approaches.

There are commercial tool management systems that provide paths for integrating database information, data transfer of information among computer systems, and software products for communicating events to other products. In fact, it would not be inconceivable to purchase one vendor's tool management system, and two years later purchase another brand and transfer information from the first to build the database for the newer, more innovative second tool management system. Then, two years after that purchase, a third system could be used from yet another vendor. The growth pattern will depend entirely on the degree of tool management that your company requires. The cost of all three systems will probably be less than the cost of either of the first two approaches.

Innovations that are occurring today may be out of your reach, so a simple tool management system may appear to be the best choice. However, two years from now, the need for scheduling tool packages (kits) and reserving those tool assemblies may be a real need. The simple system you buy today may not be capable of building tool kit assemblies to meet this need, and this deficiency may require a change in software or operation that is unreasonable to undertake. Another tool management product may address these needs in a manner that is not as difficult to adopt. The data files from one system can usually be used to create similar files, or at least assist in building the database for another tool management system. This feature or competitive edge that software vendors continually vie for benefits the end users by assuring that they will never run out of choices for third-party tool management systems.

This sophisticated tier of tool management systems already exists today. Some companies offer only one solution to tool management and it could be just what you are looking for today, but how about tomorrow? Other companies have developed a multilevel approach to tool management in which one system builds the basic requirement for the next level. This building block approach is attractive because you get (and pay for) only as much software as you need for the implementation task at hand. When the next step of tool management is needed, and the company has a module or upgrade product that fits your growth pattern, the implementation and learning curve task is greatly reduced.

What level of tool management should you start with? Most companies have a hard time just getting a tool crib's inventory under control, let alone tackling any of the following goals.

- To originate a tool requisition in the tool crib, passing it to the in-house purchasing system (on a different computer system), passing the purchase order assignment and final purchasing quantity, price, and delivery back to the tool management system.
- To check the availability of tooling required for a job or group of jobs in advance, order the shortages, and reserve the correct quantities for use when the jobs hit the shop floor.
- To accumulate the cost of tools, fixtures, and gages used on the job, scrapped by the job, reworked or repaired after use on the job, and returned as usable tools to the tool crib from the job.
- To interface directly to vendors and manufacturers of tools, their databases for tool availability, and reordering via fax, direct computer terminal emulations, and EDI techniques.
- To link to and download to presetters and other CNC controls and equipment.

These future innovations are here now. They are being used by commercially available tool management systems today. In fact, some of these techniques have been available since 1980 in commercial tool management products, and all have been in use since 1986. If these are the features and control mechanisms you are looking for, then a system with these capabilities is where you should start.

Section II

The Goals Of Tool Management

— Chapter 7 —

THE BENEFITS OF TOOL MANAGEMENT

EVALUATING THE BENEFITS

In the evaluation and justification of a tool control system, you will have several questions to answer. The following list of statements is provided as a guideline for your justification analysis. This list represents benefits that are available through the use of a tool control system, and would be desirable for any method of tool inventory control.

- Conserve manpower and minimize training requirements.
- Control losses and misplaced tools.
- Maintain timely and up-to-date information on tool usage.
- Prevent and identify tool inventory shortages.
- Improve inventory accuracy.
- Reduce space requirements and overhead.
- Reduce inventory levels and excessive purchasing.
- Reduce expediting time and piecemeal purchasing.
- Consolidate record keeping functions.
- Track tooling and its availability within the shop.
- Track tools in rework.
- Maintain a record of scrapped tools.
- Know the value of the total tool inventory and of tools in use.
- Reduce machine setup, tool return, and withdraw times.
- Pinpoint machining or overuse problems by employee, job, machine.
- Identify and track the activity of gages and fixtures.
- Identify obsolete tooling.

- Minimize improper chargeouts, pilferage, and losses.
- Take advantage of tool kitting, presetting, AS/RS storage techniques, and FMS or CELL concepts.
- Incorporate existing tool numbers and mode of operation into an automated system without making radical changes.

The function of a tool control system is to provide these benefits by allowing the functions to be easily reported, analyzed, corrected, and to allow for timely decisions concerning tools while utilizing the minimum amount of manpower and operational changes.

The control capabilities of a tool management system will allow you to reduce manufacturing costs and reduce fixed overhead charges. One of the basic elements is that of tracking tools, which allows for proper charges to jobs, departments, and work centers. Tracking also ensures that returned tooling is refurbished or otherwise properly disposed and recorded. Tracking builds confidence in the crib's ability to supply the tool when needed or to provide information concerning the tool's whereabouts and current status. This reduces tool hoarding on the shop floor, and, most importantly, sets the stage for accuracy. This basic ability provides several direct and indirect savings opportunities.

DIRECT SAVINGS CONTRIBUTIONS

Recording transactions of tool activity and tracking the normal life cycle of the tool sets the stage for accuracy. The accuracy of the tool tracking and activity records develops the information that supports and justifies the conclusions regarding tool activity cost analysis, inventory reduction, tool rationalization, and standardization.

A cost analysis of the tool's activity enables management to track manufacturing related tool costs in a timely fashion. Tracking tool costs can point out budget over-runs, bottlenecks, rework costs, unusual purchase costs, and scrap cost. Many times, this information, provided early in the part manufacture process, can make the difference between a profit or a loss. Tooling costs, in general, are drastically reduced through reduced inventory, reuse of used tools, and less expediting of tool orders. Identification of duplicate tooling and establishment of adequate quantity levels for each tool can contribute to a 25% reduction in overall tool crib inventory. Consistent tool management and tool tracking ensure that this benefit is enjoyed throughout the manufacturing process year after year.

Tool rationalization and standardization of tool purchasing, tool selection, tool usage, tool type and sizes can be established. Standardization can reduce duplication and identify potentially obsolete tooling. Standardization and tool rationalization are vital to the foundations that will allow for the expansion of your tool management efforts to include Tool Resource or Requirements Planning techniques.

INDIRECT SAVINGS CONTRIBUTIONS

Cost identification of durable and perishable tools, gages, and fixtures is easily captured through the normal activity transactions involving tools. What is usually overlooked is the cost of conducting the transaction and the labor overhead charges for supplying tooling to the shop floor for use. Most of these costs are absorbed through indirect labor budgets, included in overhead burden rates or deducted as cost of doing business expenses. A savings in these areas is generally acknowledged as beneficial but hard to recoup or identify as hard dollars. Some of the indirect savings areas that contribute to the justification of tool management are labor savings of machine operator's production time, better utilization of crib personnel, contributions to less machine downtime, and, most desired, improved part quality.

Labor savings can be realized in the time that machine operators save while waiting for tools, setups, and tool setting. Reducing this unproductive time caused by related tool delay charges can save a company over $1.500 per year in labor overhead per operator. Crib attendants recover time that was previously wasted by outdated methods of inventory control, and this time can be better spent focusing on tool inventory and issue activity using a defined procedure for tool management. Their time is now used managing, not expediting, the tool inventory, and their services can be expanded to other tasks within the tool crib.

With the appearance of the right tools at the right place at the right time, less machine downtime is realized. Tool shortages at the machine have always been a contributor to production bottlenecks, but tool availability at the machine now becomes a major contributor to reducing the intangible overhead burden rates attached to the machine. Any reduction in overhead, machine and operator burden rates, setup time, and production time help contribute additional revenue because of lower chargeable production costs. In addition to profit generation through the reduction of these charges, one additional benefit is improved quality. Better control generally results in standards being followed; procedures being monitored efficiently, and better use of tools, machines, and labor; and this helps reduce scrap and defects in manufacture.

SAVINGS SUMMARY

The following is a summary of direct inventory cost reduction and saving areas. The percentages related here can easily be achieved and are often exceeded by most tool management installations. These percentages reflect the amount of contribution recovered from the total tool inventory:

- up to 15% in the reduction of identified obsolete tools,
- up to 30% in identification of duplicate inventory,
- up to 20% return of hoarded tools,
- up to 5% in shrinkage reduction, and
- up to 30% stock reduction of excess tool levels.

Depending on how disorganized your tool crib is, if you realized a savings benefit from each of the above areas, the savings would be large enough to restock another identical crib. This may not seem to add up correctly, but if you consider the current condition of your unmanaged tool crib, you probably do not have enough organized information to dispute the potential savings represented in the above figures.

Similarly, proof of savings from indirect sources should produce the same questions. A summary of labor cost saving areas that can be realized include the elimination of:

- tool searches and expediting,
- handwritten tool receipts,
- card system,
- manual inventory functions,
- manual purchase requisitions, and
- delay times at the crib window.

Other overhead savings include:

- machine up time increase because of fewer tool shortages,
- better setup times, and
- correct tool issues.

Quality improves because of:

- less scrap and repairs, and
- the correct tools and inspection equipment are provided at the manufacture point.

CASE HISTORY EXAMPLES

Tooling is a perpetual cost to manufacturing, and very little accountability is enforced once tools are issued for use on the shop floor. Lost tooling, dull tools, and the wrong tools appear and disappear on every job in every shop on a daily basis. The availability of tool management systems has provided a means of controlling these *perishable assets* for manufacturers, and has contributed substantial savings in new tool purchases and in manufacturing costs.

If this discussion is difficult for you to relate to your actual manufacturing environment, consider these following case histories of companies that have related their tool management experiences to me.

Tool Management User 1

A provider of products to the dental industry found that their approach to CIM and their use of CAD and CAM systems was revolutionized by their tool management system. Tool designers, CNC programmers, process planners, accounting and crib personnel all share information now contained in the tool management system database. The tool database information is routinely transferred between several other in-house applications programs and manufacturing control systems — specifically, their MINX Material Requirements and Planning system, their Weber CAD/CAM system, and the in-house mainframe that handles the general accounting and purchasing functions. The manager of Manufacturing Engineering explained that "*since the tool management system tracks the activity of tooling on the shop floor, almost immediate information is available about machine performance and the progress of work-in-process. This allows us to spot bottlenecks in production before they become a serious delay. Many times the use of tooling is a better and faster indicator of a delay than the shop floor data collection system. We have organized our manufacturing operation to specify kits of tools for all operations in piece part production and have substantially reduced setup and overall production times. Kitting through the tool management system ensures that the correct tooling, fixtures and gages are used and available when needed. Our application of kitting tools for part operations has reduced our physical tool crib inventory, while with a smaller inventory we have had almost no tool shortages.*"

Tool Management User 2

Before installing their tool management system, a manufacturer of printing equipment had a constant problem with tool shortages in their tool crib. This was the primary contributor of production delays on the shop floor. Expediting rush tool orders to the purchasing department was a daily emergency. The tools were in the shop, but they could not be located when needed.

The payback for the tool management system was initially estimated to be two years. There were no real records to back up this hypothetical payback, even though they were using a computer for tool inventory control. The inventory control system did not know anything about the tool after it left the tool crib, and, since it was hard to use, many withdrawals were never entered. The existing computer system was rarely used on second- and third-shift operations, so the inventory was rarely correct. The driving criterion for justification was the manufacturing manager's gut feel, rather than factual hard numbers.

It was demanded that when the new tool management system was installed, all crib personnel on all shifts would use the system to record the activity of every tool that was issued, returned, reworked, received from vendors, or scrapped. Training of all personnel was performed in less than three days and the system went operational two months after installation. The performance and improved operation of the tool crib was far beyond expectations. There were no longer any tool shortages — in fact, tools were returning from the shop floor in such great quantities that the tool crib became so overstocked that hardly any new tools were purchased for almost six months. The perishable tooling budget for the next year was reduced by twenty percent, obsolete tooling was identified and purged from inventory, the overall inventory of the crib was reduced by ten percent and, finally, the location of all tools was no longer a secret. As it turned out, the manufacturing manager's gut feel was one and a half years off, the payback period was six months.

Tool Management User 3

A manufacturer of medical implants was trying to control their tools with a manual card system which, for the crib personnel, was impossible. The card system encompassed six tool cribs and the efforts to coordinate the personnel and the production managers they supported was difficult, to say the least. The cost of tooling required by their manufacturing techniques (many of which are secret and private), were susceptible to many price increases. It was necessary to control the usage, storage, and purchasing of these expensive tools. Manually auditing the card system for purchasing information, year-to-date, month-to-date usage, and timely reorder notification was too time consuming to provide any recommendations or order requests that were on time for tool purchases. They decided that installing a computerized tool management system would provide automatic and timely reorder of tools by tracking year-to-date, month-to-date, last year, and last month tool usage information that could guide future purchases and also track tool performance and ensure that tools would be available for use when needed on the shop floor.

One of the first notable signs that the tool management system was working was this statement from a manufacturing engineer: "*We used to send a driver out for tools on a one hundred mile round trip three times a week. In the six months that the tool management system has been on line we haven't sent the driver out for tools once.*" In addition to this improvement, through the transaction log of the tool management system, reports are now being produced that help evaluate the performance of one manufacturer's tool to another and compare the pricing of tools among suppliers. The ability to report on the peculiar activity within a manufacturing facility that specializes in special and proprietary machining process has allowed this company to spot deliveries of incorrect or lesser tooling from the originally requested tool, spot discount discrepancies in tool orders and deliveries of suppliers, and actually justify buying more expensive tooling because of performance tracking. The bottom line was not the reduction of tooling expenditures, but the reduction of tool cost in manufacturing and the increase in production and quality.

Tool Management User 4

A manufacturer of hospital operating room equipment needed to reduce the labor required to maintain tool assemblies and the number of tools required to support the availability of the assemblies for their CNC manufacturing facility. They were using a manual card system that did not allow timely identification of tools for reorder. Expediting orders for tooling was the daily routine, not the exception.

Installation of a computerized tool management system immediately provided tool inventory visibility and allowed timely ordering and stock level control. The kitting ability of the tool management system allowed control, visibility, and resupply of the tool assemblies to the CNC machines.

Other benefits that were realized after installation of the computerized tool management system included: quicker tool crib window reaction time to tool requests; better kit setups with less substitute tool changes; and an overall reduction of on-the-shop-floor-tooling. Less tooling was being squirreled away on the shop floor and more returned to the tool crib. This enabled them to resharpen tools and provide assemblies according to CNC program needs while maintaining adequate inventories to meet production requirements. The kitting control of the tool assemblies was believed to assist in the increase in the quality of piece part production.

After the first year of using the computerized tool management system, the tool crib inventory was reduced by $350,000. In addition, the setting of proper economic order quantity levels for perishable tooling reduced new tool purchases by $65,000.

We could tell user stories all day. The point is that you must understand your current tool management procedures and recognize the opportunities for improvements. Then communicate these areas of tool management opportunities to others in your organization. This is the start of the justification process. Justification studies require a champion; but to implement a tool management control systems requires a *team*.

— Chapter 8 —

JUSTIFICATION ANALYSIS

WHERE TO START

The purpose of a justification analysis is to point out the numerous benefits that can be realized by your decision to implement whatever it is that you are trying to justify to others. Generally, only a few of the most glowing benefits are actually mentioned. Many more will surface, if and when the tool management system is implemented, but they will probably go unmentioned thereafter. An effective tool management system dramatically improves tracking and availability, and it reduces inventory control problems both in the crib and on the shop floor. The benefits are so numerous that it will be difficult to document all of them. In addition, the benefits are people specific — in other words, the benefits to the crib attendant and the machinist or the shop floor foreman are important and noticed, but only to them, and hence they are less likely to be documented. But the impact of the success of the tool management system's implementation on these individuals may be as important as the enormous cost savings predicted in the formal justification document.

There are a few points of consideration you should keep in mind during your justification process. The numerous benefits and cost savings that are expected and believed achievable are common throughout the manufacturing community. However, difficulty in documenting the actual cost saving and effectiveness of a potential benefit arises when looking for the facts to base the justification. Why? Because the historical tool usage data and other information is out of date and recorded on manual cards or, even worse, buried in folklore. And the longer the tool crib has been in existence, the harder it is to find any reliable justification facts and figures. Try finding which tools are obsolete with your current record-keeping system; and if you actually produce a list of tools, does it cover all tools or only a defined set. You cannot produce an accurate cost of tooling in the tool crib from purchasing records either, unless you can produce lists of reworked tools and scrapped tools, and a value for tools on the shop floor. How will you produce the record that identifies the hours that management, shop foremen, and operators searched for tooling on the shop floor, or the cost of air freight and next day air charges for rush orders?

INTANGIBLE BENEFITS

Many of the benefits you will be looking for and basing your justification on will be intangibles. You know you are losing money in an area, and you know better tool control could prevent it, but you can't put your finger on it or document the actual costs. Without supporting evidence, it will be hard to make a case for the justification point. This is a key difference between manufacturing personnel and financial personnel. Manufacturing knows that if a machine breaks down it must be replaced to continue the production line. Finance wants to see the paperwork that says the line is going to stop before you can buy the machine. Everybody has to justify their job, even if it's just standing in each other's way. It's the lack of evidence that makes the justification so hard to prove, but the promised payback from these intangibles is so attractive that it overshadows any justification that can be shown through traditional methods. The people who fight so hard to convince others of the potential payback from intangible benefits have taken a stand and made a total commitment to the implementation of the tool management system's concept. This commitment ensures the success of the system's implementation. But what kind of unknown instinctive insight can provoke such commitment? Those who become committed to the idea of a tool management system feel they can document these paybacks if they have a system in place, and they are certain that time will prove them correct.

The control of tooling from the issue point — the tool crib window — gives the ability to reduce the unnecessary issue of perishable tools. This reduction can be measured in reduced tool purchases; but until some records can be produced, it is an intangible benefit. This will reduce consumption of perishables (and durables). The control of the issue transaction will reduce the frequent excuse of lost tooling. Since the tool issue transaction was recorded, the "*Oh, it's lost*" excuse is harder to defend; there are no lost tools: they are misplaced, hidden, hoarded, scrapped, and borrowed. Or some operators just feel "*I'm too lazy to follow your procedure of cleaning up after the job and returning the tools.*" This last reason is the unspoken truth. Most shop floor foremen do not want to enforce such a procedure because they feel it undermines their popularity.

Because tool crib transactions are being recorded, the obvious benefit is accurate inventory records. The intangible benefit is the feeling that this will allow a reduced inventory level to be maintained — yet it will still be large enough to keep the shop running. The discipline factor is the establishment of realistic minimum and maximum stock levels. This was difficult to gauge or establish in the past since there was no way of monitoring the inventory's actual activity and establish control levels. Part of the stock level difficulty is the fact that so many of the tools are stored on the shop floor, either in cabinets or in the employee's tool box. The control of these tools has been lost as soon as the record maintenance activity of in-use tool inventory is removed from the tool crib. Recording tool transactions allows the crib to regain control over the inventory, even if it is not physically within the tool crib. Eventually, all tools in and out of the crib become accountable to the controls established by the tool management system.

The largest intangible benefit that drives the need for tool management is visibility. Transaction records allow tools to become visible to management and anyone else who wants to know what activity has transpired for a tool, a given set of tools (kits), for a particular job, for an individual machine, or with an individual employee. Visibility reduces unneeded requisitioning of tools, prevents the lost-tool excuses, and provides an analysis factor when reordering. The unspoken cost justification here is reduced ordering cost. Visibility reduces searching, the cost associated with the search, and the cost of lost machining or production time during the search. These searches and production delays have a intangible cost that cannot be easily proved or documented.

The on-line ability of having the tool inventory and its location available at your fingertips gives everyone a higher degree of confidence. This feeling of manufacturing confidence provides its own intangible benefits. Whether this manufacturing confidence allows personnel to do their jobs better, or prevents misuse of tools, or reduces tool shortages, it is definitely tangible. Many companies have experienced this attitude among shop floor personnel, especially those directly involved in the management of tools. Can a dollar value be assigned to a feeling of confidence? I am sure some professor at some university can come up with a socially motivated theory that will promote better workmanship because of manufacturing confidence, but we will just say that it can't hurt.

TANGIBLE BENEFITS

While we may have difficulty assigning dollar values to the intangible, the longer the tool management system is in use, the more information it can produce to prove the value of what once was considered intangible. Tangible benefits are supposed to be supported by hard facts and provable costs. Tool management techniques in place prior to the introduction of a formal tool management system have seldom provided hard facts or traceable accounting costs. The most obvious starting point is a tool purchase history, and I will save that discussion for later in the tangible benefit analysis. The most common attempt to produce a cost justification for tool management is in direct and indirect labor savings, in lost production time, and machine spindle idle or downtime. The fact that some estimates and facts can be gathered about these delays, and a cost can be associated with them, marginally qualifies them for consideration as tangible justification benefits. In reality, at most companies they are probably intangibles. If you can't record production delays or maintain production schedules, how can you even begin to supply information where tooling is a recorded delay problem? Of course, many shops use the tool delay reason for all production bottlenecks. So this benefit is definitely a tangible to them, and justification for a tool management system is an embarrassment and the main reason they have not implemented tool control yet. However, it is of no use to continually remind you that implementing a tool management system will take away one of your most reliable excuses for why you are not meeting production. Someone is going to center around the fact that labor and machine costs are part of this justification, and a reduction of these costs is a major goal of tool management.

Lost production time due to tool delays usually is assumed to be "doing business as usual," an expected part of the everyday routine. When it happens, it is accepted, and very little comes of it after the delay is solved. More attention is directed at getting the job back on track, as it should be. The usual result is that more emphasis is spent on job, material, and labor control, but what caused the tool delay problem is never discovered and remains free to repeat itself. Even though this is really an intangible, it produces a noticeable delay and a cost can be associated to it. It is usually lumped in the overhead burden of indirect labor or other budgeted cost. These delays contribute to increased production times, inflated overheads, destroyed employee spirit (where is that university professor?), and increased costs assigned to the price of the product. The last effect is what the company is going to really feel when their competition puts its tool management system in first.

Where can you start looking to document some evidence to support a justification based on the reduction of tool delays? A starting point could be the time cards. If you have categories that allow operators to clock-in the various reason for delays or how they spend their time, you can compile a list of typical times that are tool delay related. The list of delays could include changing worn tools, tool breakage, waiting for tool setups, or searching for missing tools. If the time cards are not detailed enough, then a survey or sampling of production delays needs to be conducted. The justification section of this book includes some samples you could adopt for your survey.

Now that you have gathered the tool delay evidence, how do you use it? Formulas will follow later, but a practical example usually works better for manufacturing people. Let's say you came up with a conservative estimate of delays by machinists on their time cards that went something like this. You found from the time cards or your survey that an estimated 35 hours per month were lost because of tool related delays. And your average machine burden rate is $85 per hour. This computes to $2,975 dollars a month and further computes to $35,700 per year. Now because this amount may appear to be too much to some and not enough to others, a more in-depth study is conducted. This time the study reveals that delays logged by machine operators against production bottlenecks for other reasons actually have tool delay reasons associated with them, and there is another 10 hours per month of tool delay time that should be counted. Now you are up to $45,900 a year. You should have left well enough alone when you had the chance. It doesn't sound like a lot of money, you say. This is a yearly recurring preventable cost. It's $459,000, and more, if you count inflation, over ten years. It could be another paid employee in the tool crib or on the production floor. It's a new cutter, grinder, or machine tool. You're beginning to get the idea — it's found money. Of course, it may not be considered hard money. This is where that tangible versus intangible definition becomes important.

However, if you study the delay reporting in-depth with your shop foremen and add in the time they contribute to the solution of tool related production bottlenecks, the picture drastically changes. At a manufacturer of large machinery used in the beverage container industry, a survey revealed that over a two-shift five-day operation, 68 production hours per week could be lost to tool-related delays. This manufacturer estimated that the tool delay situation was costing them $265,200 a year. It was reasoned that if they were only 50% right, they were still losing the opportunity to save $132,600 per year. Or, more importantly, to realize an additional $132,600 in profitable production time would actually quadruple the profitability of the product.

Justification through soft dollar overhead and labor costs will meet objections. There are those who will argue that the savings will not be as readily recoupable as everyone wishes, and they may be right. I like to argue that if only 10% of these cost can be realized, then that savings alone can justify the purchase of over 75% of the commercially available tool management systems on the market today, and provide a one year or less payback, or return on investment.

However, a hard-dollars justification is always more palatable to all involved. The obvious way to prove that actual money can be saved is to follow the dollars that are currently being spent on tooling. This method also gets dangerously close to the political aspects of tool management and its opponents. Where there is money involved, there is also power, authority, and position. Sometimes revealing how the money allocated to the tool budget is spent is not welcomed by all involved in the budget approval or spending authority. Tooling budgets are real. They are tangible dollars and they are part of the financial statement. Documented savings here will be hard to refute.

As you start looking into the documentation, you might be surprised at how little of it exists. Typically, only the cost of purchasing tools can be truly traced. In other words, the purchase of most perishable tools can be readily accounted for, and the direct cost of durable tooling can be traced. The purchase of specials, modified tooling, and in-house built or designed tools is a little harder to directly account for. These costs are sometimes bundled with other costs (design or setup of a part for production) and are not clearly designated against tool budgets. Another tooling cost that sometimes eludes the audit is the cost of tooling packages purchased with machines. And yet another tooling cost is that of repaired and reconditioned tools.

Tool suppliers also figure into the cost of purchasing. That is why so many are now beginning to give you, their customer, the benefit of consolidated buying power with them and the tool manufacturer. They are offering sole source supply contracts and blanket order options on minimal tool orders that they would not have considered doing a short time ago. They know that if you start tracking the tool usage and order histories, and consolidate your tool buying, they could be in danger of losing your business to a competitor that supplies you with better insight to your tool usage and purchases. No matter what deal your suppliers offer you, the only way to correctly control your tools in use is through your own in-house tool management abilities. Ordering tools is a convenience, not a substitute for shop floor control. Reorder notification is one of the by-products of tool tracking.

The ability to track tools and inventory gives you the ability to order when supplies are below your minimum standard. The ability to reorder on notification gives you the opportunity to record the history of ordering and receipt of tools to the tool crib for replenishment. This tool control ability gives you an opportunity to evaluate a vendor or supplier's responsiveness and the supply history of tools. Couple this with the usage history and you can evaluate both the supplier, the quality of tools he supplies, and the price of the tool. This control usually results in less tool ordering and reduced tool expenditures. It also reduces the number of orders for a tool and reduces the associated paperwork cost for purchasing tools. The reduced paperwork may not be savings in tool cost, but consolidating your tool orders may allow you to lower tool costs by buying in standard lots, package sizes, or through defined blanket quantities on a schedule. Consolidated purchases can reduce the cost of the tool purchase by 5 to 7 percent or, in some cases, more. Proponents of electronic ordering systems claim that reductions in time spent, paper generated, and the cost of communicating the order provide bigger justification and pay-back on the longer term than had been anticipated when the tool management system was implemented.

The most obvious tooling to review for substantial and immediate savings are perishables, and the most common perishables are carbide inserts. Savings of 20 to 40% of perishable tooling purchases are not unreasonable. To prove my point, tell your biggest supplier of carbide that you are putting in a tool management system and will be evaluating the performance and pricing of your current carbide usage, then sit back and observe the reaction. You will get offers for special pricing, consolidation studies, integrated supply, sole source supply of any insert (from any manufacturer) you need, and even an on-site resident to issue the tooling for you. Just follow the money, and the reason for all the attention will soon become apparent, and so will the internal political ramifications. Without scaring the living daylights out of you and revealing an actual dollar amount, consider that most users of tool management systems are saving 20 to 40% (conservatively) each year on the purchase of perishable tools alone. If you saved only 10% of your yearly tooling budget, would you put tool management to work for your company? And what if the payback was within two years using this figure alone? You bet you would; and if you didn't, and word got out that you didn't take the opportunity, it should cost you your job.

Tool management, used properly, is an invaluable resource that will help identify other cost savings opportunities. Savings in tooling are real and they contribute to the third largest expense you have in producing parts. Tool control invites improvement in all aspects of production. Gaining control over tooling gives a positive information feedback opportunity to production capability and shop floor operation.

DOING THE FIGURES

The following formulas are provided to aid your quest for justification facts and figures. You will want to calculate your own figures to enter — the ones represented here are for a plant about 250 hourly people, working 2 shifts a day, about 17 CNCs, one main crib, and a pre-setting crib. Software is installed on an AS400 running MACPAC general business and consists of Material Requirements Planning programs, a couple of CAD/CAM packages (BRAVO and CADAM). There are several PC's distributed around through the office and they are manned by engineering personnel running the familiar ilk with their mice. But they don't have any formal tool control (at least, any that they can reliably gather information from). Their justification summary could easily look like the following.

Assumptions for Use In Computation Parameters
Annual Perishable Tool Budget/Spend: $1,250,000.00
Indirect Labor per Hour: $25.00
Average Machine Hour Burden: $85.00
Number of Machines on the Floor: 60

Average Savings Statistics of Experienced Tool Management Users
First Year's Inventory Reduction: 15% to 25%
Machine Downtime: 10 to 20 hours per machine per year
Recovered Man Hours, Shop Floor: 90 to 150 hours per year per man
Lost Tooling Recovery: 5 to 12% of yearly purchase

Initial Savings

One-Time Inventory Reduction $250,000
(20% × $1,250,000)
Indirect Labor Savings $9,000
(120 hrs/person/year × 3 people × $25/hr)
Recovered Production Time $76,500
(15 hrs × 60 Machines × $85)
Recovered (Lost) Tooling $106,250
(8.5% × $1,250,000)
TOTAL *Unbelievable* Projected Savings $441,750

This is the first year the system is in full operation. For the second and following years, it is estimated that $139,153 continual savings will be realized each year. This figure will increase as the cost of individual tools increases.

Required Investment

Purchased Tool Location Control Tool Management System $30,000
(documentation and warranty for one year included)
ReportWriter Option for Ad Hoc Queries $2,500
UNIX Compatible Hardware for 5 Input/User Stations $12,000
Vendor-Supplied installation and Training-10 Days $10,000
Implementation Labor— One-Year Estimate $45,000
TOTAL $99,500

Cost Justification Computations

Inventory Reduction — Perishable $140,000
($700,000/year perishable purchases × 20% reduction)

Consolidated Renegotiated Purchase Contracts $35,000
([use of blanket orders, etc.] $700,000/year × 5% reduction)

Lost Tooling Prevention $62,500
($1,250,000/yearly inventory × 5% of total purchases)

Production Uptime Increase $180,625
([reduced time looking for tools, etc.] 85 hours/week × 50 × $85/hour × 50% reduction)

Productivity Increase — Preset Area $9,375
([reduced setup time due to lost, misplaced tools]1.5 hrs/shift × 2 Shifts × $25/Hr × 5 days × 50 wks × 50% reduction)

Return On Investment (ROI) 430%
([savings divided by cost × 100] $427,500/$99,500 ×100)

Payback .... 2.79 months
([cost of system divided by monthly savings] $99,500/[$427,500/12])

As the old saying goes, figures don't lie but liars can figure. Even if these figures were *only 10%* accurate, you could still get a payback in 28 months. But until you gather your figures, you will have to keep lying to yourself. Your tool supplier hopes you never start looking at your tool inventory. If you just keep treating it as an inventory adjustment, a purchasing management task, and an estimated increase in tool expenditure budget each year, the vendors will be happy and you won't be the wiser. Purchase budgets and inventory control are not the keys to reducing tooling cost. The sooner you realize that, the sooner you'll start saving money. The solution isn't in tool scheduling, sole-source vendor supplies, and blanket orders; the solution is where the tools are used. I know that the obvious is not supposed to be the solution. The solution is supposed to be more complex than this. Not everything in this world is complex or a conspiracy — only if you allow it to be.

— Chapter 9 —

HOW TO JUSTIFY THE SYSTEM

Tooling generally represents the third most costly function in a company. As perishable tooling continues to rise in cost, it is becoming difficult to ignore the amount of dollars expended to restock the tool crib inventory and to maintain sufficient quantities to meet the continual production schedules.

When looking at tool control, one question should be asked and answered first: *What is the current value of the tools within the crib?*

As you begin to answer this question, it becomes evident that *the tool crib may only represent one-third of your tooling investment.* The other two-thirds may be stored or actively in use on the shop floor.

In most companies, a Capital Equipment Expenditure must be submitted for approval before any system purchase can be made. You must follow the procedures for budget submission, approvals, and expenditure that are required by your company. A project that concerns tooling generally attracts the attention of several departments and their managers. This will result in numerous meetings. The suggestions made in this chapter are meant to assist your effort in defining tool control guidelines during these meetings.

One subject of discussion will concern the effect that knowledge about the tool inventory, its use, and location will have on the general operation of the company. Topics will range from revamping the purchasing procedure for tools to the integration within an MRP system. Basically, every aspect will be discussed, *except for how it will assist the real-time production of parts from a shop floor management level.*

Unless you have complete autonomy in the decision process, you will have to convince others of your needs in tool control. Since tool management will become a strategic information system to the company, I suggest that an evaluation team be formed with a representative from each of the following areas of responsibility: Tool Crib Management, Purchasing, IS (Information Systems), Production Control, Quality Control, and Process Planning.

The representatives on the team should be responsible for investigating the need for tool control and for constructing two lists: the "*wish list*" and a "*real world*" list of tool control needs for their area of responsibility. Since tool control previously may not have been formally monitored, investigation methods may be difficult to implement and the results may not be easy to interpret. One of the goals of this book is to point out the possible candidates that can contribute savings through tool management.

One of the first tasks of the team is to determine the needs and parameters of tool control for the company. An easy starting point is to boil the wish list and the real world list down to size and then summarize them. A secondary task is to explore the market for available tool management systems.

The information provided here will help guide your team in determining the problem areas, proposing solutions, solidifying justifications, and selecting or designing your tool control system. Before attempting to answer the following questionnaire, a real knowledge of your existing tool control procedures is vital. You may wish to familiarize yourself with the tool crib record cards and the procedures used by the various department heads that order, design, and assign tools for production use. You should have a copy of this questionnaire (or your modified version) filled out by each member of the evaluation team. Then summarize the results of each questionnaire for the team members. The summary will allow each team member to understand the position and level of tool control knowledge of the other team members. Before you start your survey, here are a few statements retrieved from past and more formal surveys concerning tooling.

American Surveys

- 30 – 60% of a shop's tooling is somewhere on the shop floor *Lost* and *Expensed.*
- 15 – 20% of scheduled production cannot be met because *tooling is not available.*
- 40 – 60% of a foreman's time is spent *Expediting, not managing,* tooling related problems.

Industry-Stated Needs

- Tool control is *Strategic* to the entire company. It can provide timely information about the activity and status of tooling to every department.

- FMS and CELL Concept manufacturing have suffered performance setbacks because of *poor* tool management.
- JIT implementations cannot be complete until tool management techniques have been defined and implemented.

International Consensus

- Tooling is often cited as a *significant* contributor to production bottlenecks.
- Tooling is a major source of uncertainty.
- Tooling contributes to delivery delays, cost overruns and inferior product quality.

Tool Control Questionnaire

The following questions are meant to assist you in identifying problems that your present method may have. All the factors noted are common problems that evolve in direct proportion to the length of time a firm has been in existence. You may wish to add questions that are even more specific to your company's operation. This questionnaire may be copied for use within your company.

________________________ ________________________

Team Member Department

1. How accurate are your tool inventory records?
   Assign a percentage of accuracy for quantity and value of tools located:
   a. In the main tool crib ________%
   b. In the satellite cribs ________%
   c. In each department ________%
   d. In each employee's possession ________%
   e. Currently assigned to each job ________%
2. Can you quickly determine which tools and how many of each should be reordered? (________)
3. Do you maintain records for scrapped tools? (________)
4. Can you determine the value of expended tools over any given time period? (Do not count tools squirreled away by employees, and do not count by purchases only.) (________)
5. Can you easily identify obsolete tools and their $ value? (________)
6. Could a new crib attendant learn your tool locations, and issue and record keeping procedures in less than one day? (________)

7. Are you ordering new tools when sufficient quantities may already be on hand elsewhere in the shop? (_______)
8. If a tool is unavailable from the crib stock and is needed immediately, do you have to conduct a emergency physical search by machine, department, employee? (_______)
9. Do you "spot"-order or piecemeal order tools? (_______)
10. What percentage of your crib storage space is effectively utilized by cubic foot volume? _______%
11. Could your space usage be improved by random tool storage locations rather than by dedicated storage locations? (_______)
12. Do misplaced, lost, or out-of-stock tools cause delays for:
    a. Crib attendants (_______)
    b. Machine operators (_______)
    c. Department supervisors (_______)
    d. Job/machine set-ups (_______)
    e. Job compilations (_______)
    f. Presetting and kitting (_______)
13. Do employees have confidence in tool availability, or does hoarding and over-requisitioning of tools occur? (_______)
14. Are tools reground and made available for distribution in proper condition on a timely basis? (_______)
15. Can you tell by tool usage whether a job or machine is running efficiently? (_______)
16. If tooling budgets are reduced, do you have proper control to prevent hoarding of tools at the machine and avoid crib depletion? (_______)
17. Do you have N/C, CNC, FMS, transfer machines, or other sophisticated equipment that does not function efficiently without precise control of all cutting tools? (_______)
18. Can you effectively use the detailed process sheet concept for all shop areas? (_______)
19. As a basic element in process planning, can you:
    a. Pre-kit tools (_______)
    b. Accurately define tools needed (_______)
    c. Use Quick Change tooling concepts (_______)
    d. Use Quick Change fixture/die concepts (_______)

20. To what extent do the following production delays occur?

	Often	Seldom
a. Crib attendant not at window	______	______
b. Line at crib — couldn't get tool quickly	______	______
c. Crib couldn't find tool quickly	______	______
d. Crib didn't have tool (looked elsewhere)	______	______
e. Waited to have tool sharpened or set	______	______
f. Only tool available was in use elsewhere	______	______
g. Tool was locked in cabinet	______	______
h. Crib handed out wrong tool	______	______
i. Tool was in the wrong location	______	______

21. Are you considering or do you have in place an FMS or Cell manufacturing installation? (______)

22. Do you have any implemented tool inventory or control method that you consider reliable? (______)

23. Is tool usage reporting and tracking a requirement of your ISO 9000 program? (______)

24. Is gage, fixture, die, or inspection equipment tracking and usage reporting
    a: generally desired requirement (______)
    b. requirement of your ISO 9000 program (______)

25. In general, do you have an overstock or understock condition of the following tool types:

	Over	Under	Adequate
a. Tool holders	_____	_____	_______
b. Collets	_____	_____	_______
c. Cutting tools (in general)	_____	_____	_______
d. Quick Change holders	_____	_____	_______
e. Die components	_____	_____	_______
f. Tool holder spare parts	_____	_____	_______
g. Fixture spare parts	_____	_____	_______
h. Gages	_____	_____	_______

26. If you are using a full service supply vendor for your tool stocking, reporting, and costing, questions 1 through 25 still apply, plus:
    a. Are you receiving tool usage reports monthly? (______)
    b. Are you receiving usage reports by machine, job, user, or other tracking criteria? (______)
    c. Are your gages, fixtures, tool holders, and special tools included in these reports? (______)
    d. Do you and the vendor review tool performance? (______)

e. Do you and the vendor evaluate new tool technology? (_______)
f. Do you have a method to evaluate tool versus cost? (_______)
g. Do you continue to get competitive pricing? (_______)
h. Are there restrictions or penalties concerning your use or purchase of another supplier's tools? (_______)

27. Briefly describe your current procedures for:
    a. Issuing tools to the shop floor:

    b. Receiving tools from the shop floor:

    c. Restocking from receiving:

    d. Scrapping tools:

    e. Sending tools to rework:

    f. Reordering tools:

JUSTIFICATION TECHNIQUES

There are many companies that can supply formulas and justification methods to help you prove your potential tool control savings. Most of these companies are involved with the sale of tool management software, cabinets, AS/RS storage and tool-dispensing equipment and tools. The more formulas you can lay your hands on, the more information you will have to evaluate from different perspectives. I encourage you to obtain other formulas and justification techniques to augment those presented in this book. The justification techniques demonstrated in this book are simple, straightforward, and address the impact on the shop floor. They will cover savings in tool purchases, labor cost savings in man power, overhead costs, and the savings attributed to elimination of time delays. The techniques demonstrated here will provide you with the information on which to base the savings. Hopefully, this will make your justification indisputable within your company.

JUSTIFICATION BY TOOL INVENTORY

A tool inventory survey will point out potential *lost opportunity dollar savings* that could be realized by better inventory and accountability control of tooling. The primary result of the survey will be a realization of the percentage of the total inventory which is in excess of an adequate inventory. In most cases, it is impossible to state exactly what is excessive inventory. To obtain an order of magnitude, conduct a physical survey of 20 commonly used tool line items using the form and formulas that follow on the next pages.

The reference to a tool line item means all tools of a particular type or number. For example: all 1-inch-diameter taper shank jobbers drills would be considered as one (1) tool line item, but there may be 99 of them (pieces) on hand.

Note: You may not have the man power, time, or readily available information to conduct the following surveys, but you should at least conduct an informal survey to get a feeling for these very important saving areas.

Tool Inventory Survey Form

Tool Description	Qty in Crib New	Qty in Crib Used	Qty in Rework	At Machine	Total Qty	Adequate Qty	Price Each	Excess Value
TOTALS								
					A	**B**		**E**

Tool Inventory Survey Formulas

(A) Total tool inventory = ______________________________

(B) Adequate inventory = ______________________________

(C) Excess inventory total (A-B) = ______________________

(D) Excess percentage (C divided by A)×(100) = ____________

Compute the difference between each of the sample line items Total and Adequate Inventory (A-B) and multiply by their respective price to obtain an individual line item Excess Value (E). Total the 20 values to obtain the **Total Excess Value of Sample Tools** (= E).

(E) Total excess value of sample tools = __________________

Multiply (E) times the current cost of money (1 or 2 points over prime interest rate) and divide by the number of different tools (20) used in the survey to obtain the **Average Lost Opportunity Dollars Per Line Item Per Year** (= M).

(M) Lost opportunity dollars/line item/year = ______________

Assuming the percentage of excess inventory for the total tool inventory is the same as your sample, and you know the total number of line items in your tool crib inventory, multiply this total times the **Excess Percentage (D)** to give you the **Number of Line Items in Excess** (= N).

(N) Number of items in excess = _________________________

Multiply this figure (N) by the **Lost Opportunity Dollars Per Line Item Per Year** (M) to obtain the **Total Lost Opportunity Dollars Per Year**.

Total lost opportunity dollars per year = $ ______________

Your dollar value number may seem unbelievable because: 1) it is possibly real, or 2) you did not choose a representative cross section of tool line items for your 20 samples. You may wish to reduce the Excess Percentage (D) as suggested in the following example, by using 10% as the Excess Percentage.

For example: If the Excess Percentage (D) was only 10% of a 10,000 item inventory, the number of items in excess would be 1,000; and if the Lost Opportunity Dollars per Line Item (M) was $85.16, the Lost Opportunity Dollars Per Year would be: 1,000 times $85.16 or $85,160.00 per year.

No claim is made that the use of an effective tool control system will result in a direct savings to your company of this amount. Rather, effective controls of tool inventory will result in the purchase of more productive and cost-effective new tooling, without significant increases in overall tool budgets. You will be able to spend your tooling dollars more wisely, while maintaining minimum safe inventory levels, recapturing scrap value of obsolete tools, recycling tooling through rework on a timely basis, and introducing new tooling concepts.

JUSTIFICATION BY LOST TIME

Lost time due to tool delays is usually not taken into consideration. This type of delay is an intangible cost, but the delay is a contributor to the overall indirect overhead burden. In order to gain some insight as to how prevalent these delays may be, the following form could be used to conduct the survey. Delays hurt the company in production efficiency, and hurt the operators by destroying incentive. I suggest that you conduct this survey over a two-day period to obtain an average of the type of delay problems that occur. A convenient *Tool Delay Survey Chart* is provided for you.

In a survey conducted for **ITC Integrated Systems, Inc.**, in 72 observations, delays were reported in 47% of the sample, with an average delay per observation being 0.16 hours, or about 10 minutes. Each observation was made over an average period of 4 hours. Therefore, the total number of observed working hours = 72×4 or 288 hours. The total number of delays reported over this time period was 14.4 hours. Therefore, the average tool delay per hour worked is 14.4 hours divided by 288 hours = .05 hours tool delay per hour worked. For this company, in the current year, the projected number of actual production hours for machining operations is approximately 310,000 hours. The total projected lost productive time per year is, therefore, 310,000×.05 hours tool delay per hour worked, or 15,500 hours production lost to tool delays. The cost to the company surveyed of a production operator waiting or looking for a tool is an average variable burden rate of about $22.00 per hour. The total cost of tool delays of all kinds for that year will be approximately 15,500×$22.00 or, $341,000.00

It cannot be said that this amount can be eliminated. However, by applying a rational system of tool inventory control and more efficient tool crib service, it is possible that the amount could be reduced at least by 25%, or $85,250/year. This savings will be apparent as increased operator efficiency. This savings is in addition to other justified cost savings you may compute during the tool control justification project.

TOOL DELAY SURVEY

Start Time: _______ Stop Time: ______ Total Hrs.:________

Conducted By:________________________________

Date	Machine No.	Time of Day	Time Lost	Tool Description or Type	Type of Delay 1	2	3	4	5	6	7	8	9	10
				TOTALS -										

Delay Type Key

1. Crib attendant not at window.
2. Line at crib window.
3. Crib could not locate quickly.
4. Out of stock.
5. Waited to have tool sharpened or set.
6. Only tool in use elsewhere.
7. Tool locked in cabinet.
8. Crib issued wrong tool.
9. Tool in wrong location.
10. Other reasons:

____

(Write in reason)

SORTING THE FACTS

Before starting a justification and survey of your tooling activity, you may wish to obtain some dollar facts concerning your tooling expenditures so that you may relate your justification to examples in this book. Listed below is a summary form of dollar expenditures that will assist you in understanding the examples as well as other justification forms. You may have to do some research within your company to arrive at the appropriate numbers. (Be careful whose toes you step on.)

Dollar Facts

Excess inventory value: $ ____________________

Obsolete tool value: $ ____________________

Cost of machine downtime: $ ____________________/HR.

Tool inventory shrinkage/losses: $ ____________________/YR.

Dollars expended per month on new tooling: $ ____________/MO.

Dollars expended per year on new tooling: $ ____________/YR.

Average cost per manufacturing man hour: $ ____________/Man/HR.

Difference between expensed
tooling and actual purchase orders: $ ____________________/YR.

Cost of special tools and fixtures: $ ____________________/YR.

Cost of repairs to tools and fixtures: $ ____________________/YR.

Dollar value of blanket orders: $ ____________________/YR.

If you are unable to answer all the questions above, you are not alone. When these questions can be honestly answered, you will have enough facts to adequately justify a tool control system for your company and demonstrate the potential savings.

DEFINING SYSTEM SIZE REQUIREMENTS

After completing the previous questionnaires, you should now have a better idea of the scope of your tool crib, its activity level, and its inventory size. The next step is to start determining the size of the system you will need to control this activity. Before you can determine what kind of tool control you should implement, you need to have an idea of how much data or information will be collected, and what departments or personnel will be needed to provide and use this information.

The following questionnaire will assist you in selecting the appropriate system for your tool application, and in determining the location of the system components.

Hardware and Station(s) Overview

1. Number of stations required at the crib windows: ____________
2. Number of stations required by other departments: ____________
   (do not include the Tool Supervisor station)

3. List all relevant departments and the distance of each station from computer or network nodes

4. Will optional input devices be used? _________
   (i.e., bar code reader, magnetic stripe reader, remote readers, etc.)
   Description:

5. Will an AS/RS* interface and station be required? _________
   (*Automatic Storage/Retrieval System, carousel, or tool dispensers.)
   Explain number and distance to:

6. Will an interface to a CAD/CAM station be required? _________
   Explain number and distance to:

7. Will an interface to a presetter be required? _________
   Give make, model, and control name: _________________________
   Explain number and distance to:

8. Will an interface to a host computer or network
   be required? _________
   Is requirement immediate? _________
   Is requirement local or remote: _________
   What is function of data to be transmitted?

9. What is the host computer make and model? _________________
   or what is the network system make and model? _________________

10. What will be the communication method?

11. Will the software be installed on a single user
    based PC? _________
    (i.e., a MS-DOS, MS-Windows or OS/2 computer system.)

12. Will the software be installed on a Network system? ________
    (i.e., Novell, Banyan Vines, Windows NT.)
    List type: ____________________________
13. Will the software be installed on a UNIX-based system? ________
14. Is simultaneous access of the tool database required at:

	Yes	No
Tool crib attendant stations	___	___
Purchasing stations	___	___
Engineering stations	___	___
Various Manager Stations	___	___
System Administration Stations	___	___

15. Do you require record-locking on all transactions? ________
16. Would you consider file-locking only? ________

Database Size

To determine mass storage requirements, the following questions must be answered as illustrated by the example entries. The numbers shown in ( ) are examples used to derive a typical storage requirement. This will aid in determining how large and fast your hard disk drive should be, or the demand for data you will put on your network and communications.

Data Statistics Table

Number of tool items to inventory: (50,000) ______
Number of active (in use) items to track: (20,000) ______
Number of bills of tool sets (Kits): (1,000) ______
Average number of items per bill of tool sets (Kit): (5) ______
Maximum number of items to carry in scrap inventory: (1,000) ______
Maximum number of items to carry in rework inventory: (2,000) ______
Maximum number of items to carry in rework history: (2,000) ______

Minimum MS-DOS 5.0 based computer system:
For this system, the above example will need approximately 18 Mbytes of space. This includes 16 Mbytes for data files and 2 Mbytes for Operating system, application, and utility S/W. This O/S is outdated, but will still be found in use at many tool cribs. They arrive as hand-me-downs.

Minimum MS-DOS 6.2x based computer system:
For this system, the above example will need approximately 22 Mbytes

of space. This includes 16 Mbytes for data files and 6 Mbytes for Operating system, application, and utility S/W. This O/S will also be found at the root of many Windows menu managers installed on stand-alone or networked PC's.

Minimum MS-Windows (3.x or 95) based computer system:
For this system, the above example will need approximately 81 Mbytes of space. This includes 16 Mbytes for data files and 65 Mbytes for Operating system, application, and utility S/W.

Minimum typical UNIX (e.g., HPUX, AIX, SCO XENIX) based multiuser computer system:
For this system, the above example will need approximately 31 Mbytes of space. This includes 16 Mbytes for data files and 15 Mbytes for Operating system, application, and utility S/W.

The storage requirements for your system will depend upon the numbers you insert into the **Data Statistics Table**, the overhead of your Operating System, and application programs. Your operating system size may differ from those presented here. Variations in system options, revision level, utilities, devices mounted, and environment settings may support overhead sizes greater than those used in the previous examples.

QUESTIONS TO ASK THE SOFTWARE VENDORS

The following pages outline a *Software Vendor Questionnaire* that you may find useful when interviewing tool management software suppliers about their systems. I borrowed the idea from a prospect of ITC's during a bidding process for tool management software. It has been placed on separate pages so you can make copies for your personal use.

Tool Crib System Functional Requirements

__________________________ __________________________

Vendor/Supplier System/Product

The following list of requirements is for a tool crib management system and software to control the tool cribs for:

__________________________ __________________________

Company Name Location

Please address each requirement indicating if your product will meet the requirement: a) As Standard Feature, b) Available Option, c) User Modifiable, d) Vendor Modification (include any charges), or e) Not Available. If a request fits more than category, check all that could apply.

1. Perform inventory adjustment on all transactions.
   a) Standard___ b) Option___ c) User Mod.___ d) Vend. Mod. $______ e) N/A___
2. Compute and display total tool inventory value and display inventory value by tool classification, i.e., durable, perishable, gage, assembly.
   a) Standard___ b) Option___ c) User Mod.___ d) Vend. Mod. $______ e) N/A___
3. Check tool stock levels and do low quantity reorder notification.
   a) Standard___ b) Option___ c) User Mod.___ d) Vend. Mod. $______ e) N/A___
4. Print overdue on order items.
   a) Standard___ b) Option___ c) User Mod.___ d) Vend. Mod. $______ e) N/A___
5. Report and display tools currently in use on work orders (jobs), at machines, with employees.
   a) Standard___ b) Option___ c) User Mod.___ d) Vend. Mod. $______ e) N/A___
6. Produce reports showing tools used at machines, on work orders (jobs), by employees.
   a) Standard___ b) Option___ c) User Mod.___ d) Vend. Mod. $______ e) N/A___
7. Perform tool kitting with tool item availability or shortage reporting and tool item where used in kit reporting.
   a) Standard___ b) Option___ c) User Mod.___ d) Vend. Mod. $______ e) N/A___
8. Track kit items on shop floor when issued.
   a) Standard___ b) Option___ c) User Mod.___ d) Vend. Mod. $______ e) N/A___
9. Track assembly (complete kit) on shop floor when issued.
   a) Standard___ b) Option___ c) User Mod.___ d) Vend. Mod. $______ e) N/A___
10. Capable of nesting tools and kits within kits.
    a) Standard___ b) Option___ c) User Mod.___ d) Vend. Mod. $______ e) N/A___
11. Capable of tracking nested tools and kits issued to the shop floor and in storage locations.
    a) Standard___ b) Option___ c) User Mod.___ d) Vend. Mod. $______ e) N/A___

12. Capable of reserving tools for upcoming jobs (advanced need).
    a) Standard___ b) Option___ c) User Mod.___ d) Vend. Mod. $______ e) N/A___
13. Able to produce tool scrap reports by machine, job, employee, and date scrapped.
    a) Standard___ b) Option___ c) User Mod.___ d) Vend. Mod. $______ e) N/A___
14. Able to track tools and kits to issue locations (machine, job, employee), scrap and rework.
    a) Standard___ b) Option___ c) User Mod.___ d) Vend. Mod. $______ e) N/A___
15. Report all tools reserved by job number or machine number and date reserved.
    a) Standard___ b) Option___ c) User Mod.___ d) Vend. Mod. $______ e) N/A___
16. Capable of accessing tool record by tool number, description, classification, location, etc.
    a) Standard___ b) Option___ c) User Mod.___ d) Vend. Mod. $______ e) N/A___
17. Capable of accessing any information contained in the tool database for customized reporting .
    a) Standard___ b) Option___ c) User Mod.___ d) Vend. Mod. $______ e) N/A___
18. Capable of reporting tools that have not been used for a certain time period, or obsolete tools.
    a) Standard___ b) Option___ c) User Mod.___ d) Vend. Mod. $______ e) N/A___
19. Able to order tools that do not need to be ordered, or tools that are not database items yet.
    a) Standard___ b) Option___ c) User Mod.___ d) Vend. Mod. $______ e) N/A___
20. Capable of transferring on-order requests to in-house computer, other software applications or EDI control software.
    a) Standard___ b) Option___ c) User Mod.___ d) Vend. Mod. $______ e) N/A___
21. Produce purchase order reports by vendor, due date, tool item, and requisition number or purchase order.
    a) Standard___ b) Option___ c) User Mod.___ d) Vend. Mod. $______ e) N/A___
22. Capable of performing tool issue, tool return, tool scrap, tool restock, tool reissue, tool refurbish, tool return from refurbish.
    a) Standard___ b) Option___ c) User Mod.___ d) Vend. Mod. $______ e) N/A___
23. Capable of defining vendor, employee, machine, job, and tool order numbers and files.
    a) Standard___ b) Option___ c) User Mod.___ d) Vend. Mod. $______ e) N/A___
24. Provide file maintenance ability for vendors, employees, machines, jobs, tool kits, transactions.
    a) Standard___ b) Option___ c) User Mod.___ d) Vend. Mod. $______ e) N/A___
25. Able to produce cost reports by employee, machine, job and tool item status.
    a) Standard___ b) Option___ c) User Mod.___ d) Vend. Mod. $______ e) N/A___
26. Produce a listing of employees with tools signed out to them.
    a) Standard___ b) Option___ c) User Mod.___ d) Vend. Mod. $______ e) N/A___

27. Able to assign catalog or commodity code to tool items and report tools by their commodity code.
    a) Standard___ b) Option___ c) User Mod.___ d) Vend. Mod. $______ e) N/A___
28. Capable of tracking tools to repair/refurbish both in plant and with outside vendors.
    a) Standard___ b) Option___ c) User Mod.___ d) Vend. Mod. $______ e) N/A___
29. Capable of tracking tools, fixtures and gages to outside vendors or sister plants.
    a) Standard___ b) Option___ c) User Mod.___ d) Vend. Mod. $______ e) N/A___
30. Capable of recalling and tracking gages and fixtures.
    a) Standard___ b) Option___ c) User Mod.___ d) Vend. Mod. $______ e) N/A___
31. Capable of producing a gage recall inspection history with last user of gage.
    a) Standard___ b) Option___ c) User Mod.___ d) Vend. Mod. $______ e) N/A___
32. Capable of producing a complete user and inspection recall history of gages and tools.
    a) Standard___ b) Option___ c) User Mod.___ d) Vend. Mod. $______ e) N/A___
33. Capable of using bar code input and remote bar code devices.
    a) Standard___ b) Option___ c) User Mod.___ d) Vend. Mod. $______ e) N/A___
34. Capable of operating on a network without database locking when accessed by more than one tool number or report requested.
    a) Standard___ b) Option___ c) User Mod.___ d) Vend. Mod. $______ e) N/A___
35. Capable of locking certain users out of system features, i.e., only supervisor can edit tools, or create tool orders.
    a) Standard___ b) Option___ c) User Mod.___ d) Vend. Mod. $______ e) N/A___
36. Capable of redirecting reports from user's console to remote printers, local printer, screen, or system spooler.
    a) Standard___ b) Option___ c) User Mod.___ d) Vend. Mod. $______ e) N/A___

Cost of proposed product to meet above requirements: $____________
Yearly maintenance/update charges for this system: $____________
Installation/training charges for proposed system: $____________
Programming language use: ______________________________
Third party or optional products needed: ______________________________
File structure of database: ______________________________
Installed operating systems supported: ______________________________
Number of installed systems (customers): ______________________________

Include with your response any exceptions or explanations of any item questions that may need clarification of your product's ability to comply with our needs. Include any brochures or demonstration materials that may be applicable. Include a brief company history and any customer references available.

— Chapter 10 —

JUSTIFICATION ANALYSIS

TOOL CONTROL SYSTEM MODEL

This chapter describes the features of a standard tool location control (TLC) system. It will define common terminology of a tool control system and provide suggestions and models for commands, utilities, expansion considerations, and a description of some basic computer hardware that may be used.

Basic Operation

The tool control system will normally be operated on a network server or other multiuser capable environment using CRT display terminals, or PC stations using Windows 3.x, 95, 98 or a terminal emulation program located in the tool crib and supervisor's area. A tool control system should be easy to learn by the existing personnel manning the crib activity. Each inquiry of the system should be prompted by key words and lead the crib attendant through the operation by instructing the attendant to simply fill in the blanks which appear on the screen. The basic use of the system should be natural to the crib attendant's day-to-day job responsibility. It makes no difference if the operating system is MS-DOS, OS/2, UNIX, Novell, Windows NT, or some version of all the above operating through a Local Area Network (LAN).

For example, the crib attendant may be required to enter the following information requested on the monitor screen:

WITHDRAW TOOL(S)
MACHINE#:______ JOB#:______ USER#:______
TOOL #:______________ QUANTITY:______

or a supervisor may be requested to answer the following question as it appears on the monitor screen for a particular report request:

Reset (M)monthly or (Y)yearly usage counts or (B)both? ___
Do you really want to reset the monthly/yearly
usage counts (Y)yes (N)no? ___

The following section defines commands and features that are desirable for a basic tool control system. The commands are divided between *crib attendant* and *supervisor* responsibilities.

Tool Crib Attendant Commands

Certain basic commands must be available to the crib attendant to record the activity of tools and obtain information concerning their status. Through these commands, a crib attendant will receive the necessary information to issue and receive tools. The commands are as follows.

(*Note*: Commands in **parentheses** are similar or equivalent terms that are used by various suppliers of tool management products.)

WITHDRAW (ISSUE): **to issue a tool to the shop floor**. The system will request the tool number, quantity to be withdrawn, the machine, job, and user to which the tool is to be assigned. After the information is entered, error checked, and verified, the system will report the tool's assigned crib location.

RETURN (RESTOCK): **to accept a tool being returned from the shop floor**. The system will request the tool number, quantity, and machine number from which the tool is being returned. If needed to uniquely identify the specific tool, it will further request the job and user to which the tool was originally charged. The system will then ask if the tool is to be returned to the crib, to be scrapped, or to be sent to rework. After this information is entered, error checked, and verified, the system will report the tool's assigned crib location, or disposition in the case of scrapping or being sent to rework.

FIND: to locate a specific tool. After the attendant has entered a valid tool number, the system will respond with a list of all locations in which the tool can be found. This command is useful for locating requested tools, gages, or fixtures when an insufficient quantity is available from the crib.

AUXILIARY TOOL NUMBER: is used by the crib attendant when the tool number is unknown. The correct tool number can be found by specifying the type of tool (drill, endmill, tap, etc.) and two-dimensional descriptors such as diameter and length. This command can also be used by the supervisor when looking for similar tools or tool types.

RETURN FROM REWORK (REMANUFACTURE): **is used to inform the system that this tool is being returned from the rework area**. The system will ask if the tool is to be assigned a new tool number because of possible alteration during rework. For example, an endmill that has changed in diameter could be reclassified with a different tool number.

INITIAL ADD: allows the crib attendant to adjust the inventory count of a particular tool and/or change its assigned location.

ADD (RESTOCK NEW ITEMS): **allows the crib attendant to enter or restock inventory items from receiving into the tool crib.** The system updates the inventory quantity and reports the storage location to crib attendant.

USER STATUS: allows the crib attendant to display, on his monitor, the current tools that a specific user currently has charged out.

Supervisory Commands

Certain basic commands must be available which allow access to information concerning the activity of tools that is of importance to supervisory personnel. The following is a brief explanation of the commands.

ADD (DEFINE NEW TOOL NUMBERS): **to enter new tool numbers into the tool database.** The tool control system will first request that the user enter the tool number to be added. If an existing record cannot be found, it is assumed that the tool is new to the crib and the supervisor will be requested to enter the tool description and other essential parameters, including the assigned storage location, if known.

EDIT TOOLS: to modify or change information contained in the database concerning a tool number.

DEFINE KIT/TOOL ASSEMBLY (BILL OF TOOLS): **to define a list of tools or kit numbers** to be used or withdrawn/returned by a common tool list or part number/operation sequence. A Kit List can consist of tool numbers, kit numbers, or a combination. Kits can be nested (called within a kit) to an infinite level. Comment lines are provided for each kit to allow instructions or comments about the kit to be entered, as well as an additional comment line per tool number that is called by the kit. The comments provide an additional area to enter special setup information, feeds and speeds, preset dimensions, and information.

EDIT KIT/ASSEMBLY: allows all information in a kit to be changed, modified, or copied.

DELETE TOOL: allows a tool number to be deleted from the database. All references to the tool number will be deleted except from kit lists, rework inventory, and scrap files.

TOOL ORDER: to notify the system that a tool has been placed on order. The command accepts and records the tool number as on order. This command is used for entry and reporting of ordered tools and removes the tool number from the RE-ORDER REPORT. To print tools that have been ordered, use the ON-ORDER REPORT.

PURGE REWORK HISTORY & SCRAP: allows the rework history and scrap files to be purged. Periodically these files may need to be reduced in size to maintain disc storage room.

ACTIVE INVENTORY REVIEW (OPEN TRANSACTIONS): **to review on the monitor screen all active tool records**. This information may be printed by using the **Active Inventory Report** command.

TOTAL INVENTORY REVIEW: to review on the monitor screen all crib inventory records. This command is the same as the Active Inventory Review except that only the inventory records of the crib are displayed. This information may be printed by using the Total Inventory Report command.

RESET USAGE COUNTS: is used at the beginning of each month or year to reset the item activity counts that have been accumulated over the previous month or year. Totals are accumulated by "last year," "last month," "month to date," "and year to date."

Basic Reporting Functions

TOOL STATUS (OPEN TRANSACTION BY TOOL #): **to report a tool's current status**. This command reports all locations to which the tool is currently assigned, both in the crib and on the shop floor. Information concerning the quantity, machine, job, user to which the tool has been charged, and total quantity and total value is printed.

MACHINE STATUS (OPEN TRANSACTION BY MACHINE LOCATION): **to report tools charged to a specific machine number**. This command reports all tools currently charged to a given machine. Information concerning the quantity, job, user, and date the tools were issued.

JOB STATUS (OPEN TRANSACTION BY PART# OR WORK ORDER): **to report tools charged to a specific job number**. This command reports all tools currently charged to a given job. Information concerning the quantity, machine, user, and date the tools were issued.

USER STATUS (OPEN TRANSACTION BY EMPLOYEE): **to report tools charged to a specific user**. This command reports all tools currently charged to a given user. Information concerning the quantity, machine, job number, and date the tools were issued.

RE-ORDER REPORT: to report tools that have reached their reorder level. This command reports all tools for which the quantities on hand have reached their defined minimum reorder level.

AUXILIARY TOOL NUMBER: can be assigned in addition to your standard tool number for all tools, gages, fixtures, etc.. This number consists of a four character class field and two dimensional fields. It is useful for finding similar tools. A report can be generated to list all tools by their class assignment or any combination of class and dimensions. For example: a report could be requested to list all endmills in inventory or all endmills that are 0.5 inches in diameter or all endmills that are 0.5 inches in diameter and 4 inches of flute length.

OUTDATED TOOL REPORT: to report unused or obsolete tools. This command reports all tools which have not been used since a specified date. After the supervisor has entered the date of interest, the system will respond with a report containing the tool numbers, descriptions, quantities on-hand, last used dates, and dollar value.

KIT TOOL LIST (BILL OF MATERIALS, BILL OF TOOLS): **allows the supervisor to examine a list of tools in a kit or a list of kits**. This will cause the system to refer to the tools contained in the list conveniently by the assigned kit number.

KIT AVAILABILITY: prints a detailed list of each item contained in a specific kit, together with its description, level quantity required, total requested quantity required, the crib quantity, and the item shortages. **The main purpose of the Kit Availability Report is to point out shortages**.

WHERE USED: prints a listing of each kit (by kit number) which contains the specified tool number. A description of the kit and the quantity required within each kit where the specified tool appears will be printed. May also be used to find where nested kits are used.

STOCK MAP (LOCATION STATUS): **provides a listing of all storage locations within the crib** and the quantity of the tools stored in the location.

SCRAP REPORT: **reports all tools that have been returned as scrap** by machine, job, user, quantity, date, and value.

REWORK HISTORY: reports all tools that have been reworked by machine, job, user, rework department, and date.

REWORK INVENTORY: **reports all tools that are currently in rework** by quantity and assigned rework department.

OVERSTOCKED TOOLS: **reports all tools that exceed the maximum inventory level.**

TOOL LOCATION: is a summary report of all tools by description, tool number, primary and secondary/used location.

ON-ORDER: reports tools that have been placed on order and have not been received by the crib attendant. Variations of the on-order report include past due orders, open orders, and completed orders; and the report can be sorted by vendor code, tool number, requisition/P.O. number, or date.

ACTIVE INVENTORY: prints all tools currently in use on the shop floor by their tracking categories, date issued, and quantity.

ITEM ACTIVITY: prints YTD, LY, MTD, LM usage for all tools in crib inventory.

TOTAL INVENTORY: report reflects the current inventory status and general information of all tool numbers defined to the tool control system. Report also reflects quantities in the crib and in use on the shop floor.

Utility Commands

Periodically, it will be necessary to update the tool information stored on the system hard disc. In order to allow changes to the data stored, several utility procedures are usually built into the tool management system. These utilities generally appear on the Supervisor's Menu and allow changes to be made concerning all information that is entered during an ADD command entry, deletion of tool numbers, purging of rework and scrap report files, defining and editing of tool kits.

It is common to see database correction utilities and error checking utilities. These utilities usually repair counts that may have been corrupted because of careless editing or mass edit mistakes or repair file integrity because of mechanical or electrical computer and power malfunctions. To aid integration techniques, it is becoming more common to see application programs provide ASCII file load and unload utilities. These utilities all aid in the maintenance, upkeep, integrity, and data exchange control of your tool database and files.

Procedures for disc copy duplication, backups, and other operating system functions are usually supplied in the tool management system's administration menu. System administration functions can vary depending on the operating system you have chosen for your computer system. You should familiarize yourself with recommended housekeeping functions for your computer's operating system.

Expansion Considerations

A tool control system should be designed to be flexible in use and reporting. A commercially standard system should provide complete updating and transportability to future computer equipment. By standardizing the system, all future enhancements and option packages can be easily added, as required, to all customer systems now and in the future. Refer to the Integrated Tools Management diagram (Figure 10.1) as an example of planned expansion capability.

The following expansion considerations are usually the next desirable steps once a basic tool control system has been established within the company.

GAGE/FIXTURE CONTROL: allows tracking and inventory of gages, fixtures, or inspection equipment, and monitors the frequency of calibration and reports calibration times and history.

DAILY TRANSACTION REPORTING: a file that records the activity of tool transactions within the tool control system.

REFERENCE FILES: these files allow the user to maintain a list of vendor addresses, alternate tool numbers, machine locations, user names, jobs or work orders, and the accumulation of tool costs to jobs.

RESERVE/TRANSFER: allows the reserving of tools, and the transferring of tools from preset, rework, or between machines, jobs, users.

ADVANCED KIT CONTROL: provides more in-depth tracking of components contained within a kit, whether it is in use on the floor or part of an assembled kit in storage or preset.

PRESET INTERFACING: the ability to communicate and share tool kit definitions with presetting information from the tool control system directly or downloaded to the presetting equipment. This ability usually includes software to communicate the setting information to DNC files for postprocessing.

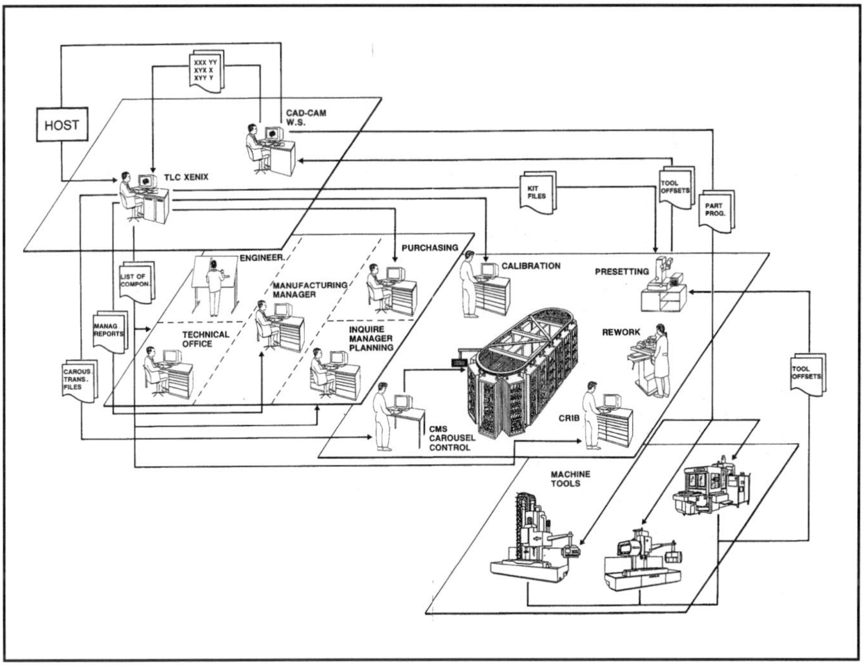

Figure 10.1—Tool Management Inter-Department Overview. *(Courtesy of Susta.)*

INTEGRATED PURCHASING: the ability to provide internal and external file transfer of ordering, purchasing, or requisitioning information of tool items. The abilities usually provide data exchange links to EDI systems or file transfer and send abilities to computer or facsimile equipment.

TYPICAL TLC SYSTEM CONFIGURATION

All tool management systems differ from customer to customer because of intended use, number of users, etc. However, they all typically have: 1) a supervisor station and report printer located in the tool crib superintendent's office; and 2) stations with ticket printers located at the issue and return windows of the tool crib.

Common additions to this configuration are: 1) a station for process planning or the N/C programming group for tool kit definition and tool availability reporting, 2) a station for tool design for the addition of new tools and the management of special tools, 3) a station or interface to purchasing for final approval or confirmation of tool requisitions, and 4) an AS/RS or carousel interface controller and stations for retrieval and pick list queuing of tools and kits. There may also be requirements for stations in rework and preset areas.

Hardware Accessories

There are several accessories that you may wish to add on, such as bar code readers or inductive chip readers, but typically the considerations are hardware options that can increase printing speeds and mass memory storage capacities. Some guidelines that may help influence your selections are discussed below. This discussion is for those who plan to implement a dedicated system for tool control. If your plan is to implement on to an existing hardware platform or network server, you IS manager can advise you as to what abilities exist and what level of expansion can be conducted.

Optional printers can be selected to increase printing speed from 120 cps to 350 cps and higher. Examples of the printing speed differences between printer speeds would be as follows, for a typical 24,000 item Total Inventory Status Report:

120 cps printer — requires 6.4 hours to print report
160 cps printer — requires 4.2 hours to print report
200 cps printer — requires 3.3 hours to print report
350 cps printer — requires 1.9 hours to print report
6 ppm laser printer — requires 1.2 hours to print report

Your choice of printers will be influenced by how often you wish to generate the Total Inventory Status Report, and by the number of inventory items it contains.

Another hardware option that adds a speed and storage capacity advantage for database purposes is the size of the hard disc drive. Hard disc drives come in different sizes and access speeds. The size you select may be influenced by the tool data storage need you anticipate or for increased growth to other application packages, such as Maintenance Management Control or FMS-related software options. The number of items you wish to inventory now and in the future will influence your choice of mass storage disc units. The following is a summary of the capacities in 15 Mb increments.

15Mb of disc — stores approximately 40,000 items

each additional 15Mb — stores approximately 55,000 items.

Today, hard disc drives range in size capacity from a minimum of 200 Mb, to multigigabyte drives that are available for network servers and stand-alone PC's.

HOST INTERFACING

Almost any computer can communicate with any other computer. The hardware to conduct the various communications modes you may desire or require is typically provided by either the manufacturer of the computer system or a third party communications supplier. The most common communication interfaces include direct dial-up modems on serial lines, Ethernet systems, and add-on TCP/IP interface boards.

No matter what hardware solution you have selected, the general question is how to get the information from one computer to another, or how to get the information from one program to another. Integration and "open Systems architecture" are the buzz words and desired goals. Today's software, application programs, and operating systems all contain and offer common data transfer utilities or tools. A discussion between your software vendors, IS department, and communications providers should define the most compatible avenue of data transfer and system integration available for your system.

To communicate from one computer to another, the communication supplier will have the right suggestion for you. An economical method is to use the public domain program "KERMIT" for simple file transfer of information between computer systems. There are also several suppliers of commercially available software for transmitting information between computers.

For transmitting information between programs or other databases you may need the help of your IS department and personnel familiar with the application programs software and database layout. Many suppliers provide their customers a set of file load/unload utilities that allow database files to be dumped to and from ASCII flat files. These utilities are usually provided at no charge and can be used for moving information from the tool control system to purchasing systems, MRP systems, CAPP packages, etc. — the list of applications can go on and on. Also available are programs that allow you to create your own reports directly from your tool control database. Generally, you have the ability to display, build, and print reports or transfer information from your database to MRP, purchasing, or other software applications.

INSTALLATION

If you are not familiar with your hardware, you may wish to have your hardware/software supplier assist you during installation. Most suppliers can provide a systems engineer to supervise the hardware installation and software checkout. This section will give you a few commonsense guidelines for installing your computer equipment.

Provisions should be made and agreed upon *in advance of installation* for remote station locations and necessary cabling, and terminal emulation software if the stations are to be personal computers. Prior to installation, final decisions should be made in order to determine the system's site location, condition, and necessary power requirements.

The multiuser computer or Network processor unit you are connecting the terminal stations to should be located in a "normal office" environment, not on the shop floor, this is primarily for the protection of the disc storage units. This area should be heated and air conditioned to avoid extreme temperature variations; however, nothing more than what normally would be required for office workers is necessary. This area should also be free of airborne dust, oils, or chemicals.

The attendant stations, unless they are PC workstations, do not require the same environmental care as the main processor. Air conditioning and heating requirements are not generally needed other than to maintain a temperature range of between 104° F (40° C) and 41° F (5° C). These requirements are generally compatible with the environment in most tool cribs.

Typically, the power requirements for most AT compatible, 80386 (rarely used today), 80486 (fading fast, but still in use), or Pentium (typical selection for small networks), systems are 115 volt, 60 cycle, 15 amp dedicated and grounded outlets of the following number: three (3) at the main processor site, two (2) at each of the remote attendant stations. An uninterruptable power supply (UPS) is strongly recommended to protect the main processor from electrical interruption or damage.

For more information, see the "Hardware Considerations" section below.

TRAINING

Your tool control system software supplier should be able to conduct applications training sessions for your personnel. Training schedules should be made in advance and should include both your crib attendants and supervisory personnel. Typically, training will consist of use of the software commands and features, diagnostic procedures, utility procedures, and hardware operation along with the other requirements that may be required for your system options. Total duration of training sessions is generally not more than three days.

Training of the systems administration personnel should be provided by the computer operating system manufacturer or their recommended training facility. Training about the operating system will prepare the systems administration personnel to support the backup, security, hardware interfacing, and communications settings that must be enabled from time to time.

There are many tool control systems in use today that employ many of the aforementioned features. The best way to evaluate the effectiveness of these capabilities is to witness them first hand.

Visit several users of tool management systems, and observe how effectively (or not) they are using their implementation of these features.

HARDWARE CONSIDERATIONS

Described in the following paragraphs are descriptions of hardware suitable for a multiuser tool control system. You can run tool management systems on MS-DOS and Windows 3.x, 95, 98 based personal computers, but most integrated tool control systems are attached through a network, Windows NT, or UNIX-based server. The personal computers then become terminals to the network or UNIX system, and are generally communicating to the tool control application through a network connection. If your need is for a single user-based system or a shared use of the application software on an MS-DOS or Windows based stand-alone PC, then most any computer hardware that can run MS-DOS/Windows and communicate to the network should have no problem running the tool control application in a single user protected mode.

But for those of you exploring the strategic sharing of tool information, you need to consider a multiuser, multitasking, record locking version of any tool control software you are going to evaluate or install. There are several hardware and communication (network) system solutions that provide adequate facilities to run the tool control application software. The following is meant to give you a brief introduction.

One solution is a microcomputer with a multiuser, multitasking operating system — in the old days, an AT compatible 80286 or 80386 processor with a XENIX or UNIX operating system was the low cost, easy-to-install solution. This is small, by today's standards, yet powerful enough to handle multiple cribs and 2 to 16 terminal stations, and it allowed many shops to enter the UNIX multitasking environment at a lower cost than networks. Today, it's basically the same but the computers are 80486/75 MHz minimum or 133-300+ MHz or faster Pentiums on the higher end. The minicomputers have become very capable in memory (8 Mb to 96+ Mb RAM) with huge disk drives (200 MB minimums to 6.4+ Gigabyte maximums).

Today's network system's capabilities almost always include some type of Windows NT or UNIX compatibility. This easily provides the multiuser and multitasking needs of large database application software, such as, a tool control application. And almost all computer manufacturers are offering a UNIX-compatible operating system. For example, IBM offers AIX, Hewlett Packard offers HPUX, SUN offers SUN O/S and Solaris, DEC offers ULTRIX and DEC UNIX, Santa Cruz Operations offers UNIX and XENIX, and Microsoft offers Windows NT. A degree of multiuser capability is offered on network servers using products such as Novell, Banyan Vines, 3 COM, or Lantastic. I know some readers would add Windows NT, but I included it with UNIX since it has been referred to as being a "UNIX wannabe."

Included within the basic system should be a report printer, interfaces for remote CRT terminals or connected PC's, a floppy diskette and cartridge tape drive, and a hard disc drive large enough for storage of the tool inventory database.

Attendant stations or remote user stations, consisting of CRT terminals or PC stations, and ticket printer combinations are added to the system as needed to allow multiple access to the tool database information.

BAR CODING AND TOOL MANAGEMENT

Bar codes have been explored as a solution for automating the transaction required at the tool crib window with tool management systems and other manufacturing control information systems. There are many devices that provide this facility. They include remote hand held units for batch operation, or wedge and RF Data Terminal units for interactive on-line operation. They may also be wand or laser readers. Bar codes cannot just be added to the tool management process. They require a planned implementation on your part.

First, you must decide what is to be bar coded and why. You also need to know when to use the bar code. You need to define the physical use and production of the bar code itself, and determine what type of bar code equipment best suits your needs.

The first candidate for a bar code is the tool. It is sometimes difficult or impractical to attach the bar code to the tool. Some tools are too small for the bar code label, some tools cost less than the bar code labeling process (laser etched), and some tools by their use will destroy the bar code. If all the tools in the tool crib cannot be bar coded, the question arises as to the viability of using the bar code on the tool.

A common solution to this problem is to bar code the storage location. This raises another question — if you do not know where the tool is stored, how can you use the bar code to issue or return it? There have been many unreasonable suggestions to this problem. For example, a catalog of the tools at the crib window with the bar code next to the tool number. Now let's think this through. The catalog will be quite thick with 3,000 to 60,000 or more tool numbers. If you cannot quickly identify the tool number needed, it may actually slow the issue process down. Flipping through the catalog for the correct tool number can be very time consuming. Some think this is simple, but it can eliminate new tools or provide access to obsolete or deleted tools if the catalog is not continually kept up to date.

Another problem arises when the tool selected is not available. If the bar code is not used as an input to the tool management system, then there is no immediate acknowledgment of the tool's availability, location, or inventory status. You may have to actually find a duplicate tool when there is an insufficient quantity available for the request — just to record the bar code.

In addition to bar coding the tool involved in the tool crib transaction, the employee is a viable candidate for data capture. Now the employee may already have a bar coded badge. If the badge is magnetic stripe coded, then you need to use magnetic stripe readers along with your bar code readers. In any event, capturing the employee number at the issue or return point is a simple process. Herein lies the solution to bar coding the tool number. If the employee could bring the tool number bar code with him, it would become a simple input. This can be accomplished easily if you have a work order or process router that could accompany each machining operation. This information can come from a Computer Aided Process Planning (CAPP) system and could provide a means for listing the tools required for the operation on the router. This solution assumes you have a way of communicating the tool numbers to and from the CAPP system. Kitting tools for the production operation can provide a simplified number that could be used in place of the entire list of tools on the router.

The more systems that contain tooling information, the greater your possibility of using bar codes at the tool crib. As you identify the tooling transaction for bar code input, you will most likely identify additional information to accompany the transaction. Good candidates for additional transaction information include the machine or cost center, and the work order or job number where the tools are used.

Once you have identified what, when, where, and why you are going to use the bar code as an input device for tool crib transactions, you need to contact a bar code manufacturer and review his products with your perceived application. Selecting the right bar code input units, label types, printing units, bar code generation software and what information is to be encoded on the label will be facilitated by the supplier of your bar code equipment.

FUTURE EXPANSION

Commercial tool management systems are continually in a state of change as new enhancements are being added. The expansion features described earlier illustrate additional capabilities that are available to further enhance and meet your specific tool control needs. In addition to standard option packages, development and research is being continually conducted for tool control needs of future enhancements such as the following.

- Interfacing to FMS control packages for tool requirement planning and tool life tracking. Advanced kit control capabilities are a prerequisite for responding to the control and information needs of an FMS system.
- Interfacing to computer controlled presetting equipmcnt for gage length and diameter setting information directly from kit description downloading.
- Interfacing to Preventive Maintenance, Production Planning, and Shop Floor Monitoring software packages.
- Interfacing to CAD/CAM systems for downloading of drawings of tools, gages, fixtures, and kit assemblies.
- Interface routines to host computers for the communication of tool kit lists, tool availability, addition of new tool numbers, and the transfer of data to and from MRP or other MIS software systems.
- Communication and information exchange to tool rationalization and tool resource planning programs.

Section III

Independent Tool Code Standard™

TABLE OF CONTENTS

INTRODUCTION TO SECTION III

Be forewarned — this document may contain errors. Most, we hope, are only typing errors, but a few may be conceptual errors. For instance, it has already been pointed out that an accepted standard exists for burrs, and this document includes its own numbering scheme for burrs. Through your comments and suggestions, the ideas presented within can be corrected, or can grow and be replaced by newer and better ideas. This is only the first revision of the Independent Tool Code Standard, and we hope that future editions will continue to improve and help bring order to an important area of manufacturing where substantial savings can be achieved.

The numbering system presented herein is not perfect, but if it helps you get your tools organized, or just gives you a head start, then it has served its purpose. Just as production of parts depends on the machines used, and the machines depend on the tools supplied to them, a tool management system depends on the validity of the tool information entered into it. The tool number is the key to the tool management records; and agreement on tool numbering is the first of many disciplines you must establish for a tool management system to be successful at your company.

— Chapter 11 —

DEVELOPMENT OF ITCS

The information used to build the basics for the Independent Tool Code Standard (ITCS) tool numbers was gathered from several sources. All considerations were based upon how a commercially standard type and size tool is made and what makes it a commercially standard tool.

The tool numbering technique described herein was developed to be practical from the user's point of view, and to be easy to enter into a computer system. Computers do not typically handle numbers and alphabetical sorts the way humans sometimes expect them to. Also, computers are unforgiving of typing errors, treat spaces and dashes as characters, and do not look at the form of the data or tool number as we do. No matter what sorting routine is used, or what format the tool number is put into, it will not mean the same thing to all the humans involved. We are individuals, and we think, look at, and interpret things differently from each other. Because of this conflict in human and computer interpretation of the same information, an attempt has been made to humanize the ITCS tool numbering technique as much as can be permitted by the software and computer disciplines that currently exist.

Tool crib inventories are typically much larger in individual items than parts inventories; tools are generally referred to by name more often than by a part number; and tools are harder to stick a label on or to permanently mark for identification. These idiosyncrasies, and many more, make managing the inventory of a tool crib far more difficult than managing the inventory in a parts crib or warehouse. Many of these problem areas are handled logically and intelligently by a competent system designed for tooling. One component of any tool control system must be agreed to and implemented from the start, namely, what we humans and the computer are going to use to identify the tools we wish to control.

Generally, the computer will go along with whatever we decide to use. So, what's the stumbling block? The human side is the stumbling block. Why? Because purchasing needs a number with which to order the tool, and the vendor needs a number to know what tool is being ordered and what has to be delivered. Engineering has other ideas about what to call tools; tool design also makes up numbers for tools; while CNC/NC programmers look at tools as complete tool holder assemblies, etc. A decision must be made as to which group is the most important, most deserving, and has the most direct responsibility for the use, performance, and handling of the tools. Actually, none of the aforementioned fits these criteria. They may call a $^1/_2$" jobber's twist drill an EDP77935, or ST-10031, or $^1/_2$" drill 50 CAT Tool Holder Assembly, or DR0500287-001, or even TD1/2. Now, with all these different ways to identify the same tool, within the same company, how can we ever expect to get agreement on a standard numbering scheme? Then to top it all off, the person who uses the tool comes up to the crib and says, "Hey, I need a $^1/_2$" jobber's drill."

The ITCS scheme looks at the way in which the user of the tool makes the request, the way the tool is manufactured, the standards that are inherent in the design and function of the tool, and the need to reference it conveniently within the computer.

Let's take a look at the $^1/_2$" jobber's twist drill that our machinist at the crib window wants. He wants a common $^1/_2$" jobber's length twist drill, not one made of cobalt, or left hand, or with high spiral flute design. What do we know about a standard $^1/_2$" jobber's twist drill? We know it is provided standard from the manufacturer with a $^1/_2$" diameter, straight shank of the same diameter, has 118 degree tool point angle, is right hand cutting, is made of high speed steel, has $4^1/_2$" of flute length, is 6" in over-all length, has a split point geometry, and costs about $7.00 for a decent one. We also probably know the vendor and some other information about it that may or may not be important at the moment. And, by the way, what is important at the moment is a key ingredient to a successful tool numbering system implementation.

Now, how would you code all that information into a tool number? Before you answer this question, ask the following question. Why would you want to code all that information if it is all standard to begin with? Because you may not have known all that. On the other hand, the machinist who wants the drill probably knows all that information; the process planner that selected it for the machining operation should know that; and the manufacturer who built it knew it. So what our machinist is asking for is just a good old general-purpose $^1/_2$" jobber's twist drill or a twist drill $^1/_2$", or a $^1/_2$" drill, etc. For reasons that will become more apparent with experience and with handling large tool databases, you will see why the ITCS code for this tool is TD1/2. If our machinist had wanted one for drilling brass, he would have said "give me a 1/2" jobber's drill for brass or with a low spiral (helix)," or a TD1/2LS.

Practical Considerations

In order to allow any tool management system to be used properly, the crib attendants and the machinist must agree to record the issues and returns of the tool they are using. If they don't, then all your inventory counts are wrong and your biggest justifications (reduced inventory, a realistic tooling budget and effective timely reordering) are nothing but myths.

ITCS attempts to code tools in manuonics — manufacturing phonics — i.e., TD for Twist Drill or CD for Center Drill. ITCS takes advantage of standards, such as ANSI or ISO wherever possible; for example, with inserts (TPG322), insert tool holders, and brazed tool bits. ITCS only adds modifiers or descriptors to the tool number to differentiate between possible duplicates and when the tool is modified from the standard; for instance, a heavy-duty 1/2" drill would be TD1/2HD. The HD modifier tells us that it has a 135 degree split point tool point angle. ITCS attempts to keep the tool number as short as possible. This is so that the crib attendant can enter it more easily on a keyboard, its construction is easier to remember, it will fit on printed reports more easily, it is easy to read, and, if bar coded, it produces a smaller label that is less prone to misreads. The basic goals of an ITCS tool number are to be as short as possible, easy to read, and easy to understand, while providing a logical grouping of similar tool classes.

An additional feature has been included with the ITCS tool number. It is the Auxiliary (Aux.) Code feature provided with ITC's TLC tool management software. The Auxiliary Code allows you to find a particular tool when you do not know the actual tool number. Inherent in its format and construction is a four character prefix, or class code, to group similar tool types, i.e., TD for all Twist Drills or TDTS for all Twist Drills with Taper Shanks. The Aux. Code also allows two major dimensions to be specified about the tool to further distinguish it within its class code. For instance, with TD class tools, the first dimension could be its cutting diameter, and the second dimension its flute length. Our TD1/2's Aux. Code would be Aux. I.D.: TD; Dimension 1 (D/1): 0.5; Dimension 2 (D/2): 4.5. A TD1/2TS would be Aux. I.D.: TDTS; D/1: 0.5; D/2: 4.375. *Whenever the Aux. Code is used to select a tool, the tool number and description are presented with the Aux. Code in order that the selection of a particular tool number can easily be made between the possible alternates that have the same Aux. I.D. and dimensions, for example, our taper shank drill would appear as follows:*

Tool No.	Description	Aux.	D/1	D/2
TD1/2TS	1/2D.,4.38FL,8OAL,#2TAPER SHANK	TDTS	0.500	4.375

The Aux. Code technique is an exclusive feature of ITCS and TLC, and is extremely useful to all departments involved with tool management. Some common applications of the Aux. Code are:

- by crib attendants when the tool number is not known by the person requesting the tool;
- when process planners or CNC/NC programmers are looking for the availability or a tool to fit a particular machining operation;
- when you want to see what tools are available of a particular class, type, or size;
- when designing new tools, to see if one already exists or if there are similar tools that could be used instead of designing a new one.

The Aux. Codes become an extremely useful tool when your database becomes larger.

Another basic need in tool management is to communicate reorder information between the tool crib and your purchasing group. Generally, the tool number scheme that works on the shop floor won't work for the purchasing department, because the ordering numbers are different than the shop's tool number. Specific and standard features of tool management systems can be useful when building your tool database and allow you to enter the ordering number within the tool record. This ability could be implemented in three possible methods where another number could be used to reference the tool during reorder circumstances: 1) to include the number in the description, 2) include the number in the vendor code, and 3) to assign an A.K.A., or alias, number to the tools database number. Please refer to Figure 11.1 for examples of these techniques.

TOOL INFORMATION

Tool: TD1/2TS

Tool Description: 1/2,4.38FL,8OAL,#2TS **EDP-123456** Cost: 8.0000
Type: (P)erishable (D)urable (G)age (A)ssembly: P
(R)eturnable (N)on-returnable: R

Reorder Point Qty: 6   Max. on Hand Qty: 24
Active Qty: 13   Quantity to Order: 12
Primary/New Location: A1C1D4C5 Qty: 24
Backup/Used Location: A11C2D4 Qty: 12

Auxiliary I.D.: TDTS D/1: .500000 D/2: 4.37500

Vendor Code: CLEVELAND**#123456** Last Used Date: 25FEB98

Qty YTD: 32 Qty MTD:11 Qty Last Year: 388 Qty Last Month: 21
Alias Number: EDP-1234567890-ABC
Alias Description: Johnson Extended Length Taper Shanks

Figure 11.1—Example tool order numbering.

I. Drilling Tools

Putting holes in metal is the most common of all machining operations, and drilling is the most widely used method of making a hole. When selecting a drill, one starts by looking at the hole diameter desired, the depth required, the tolerances to be held first, then the other machining factors.

Twist drills are the most common hole-making tools. They are made by the millions from a wide range of materials (HSS to solid carbide); come in a variety of styles and sizes (0.0059" to $3^1/_2$"); are specified in inch, metric, wire gage number, letter, and fractions; and can commonly be used by almost any machine (except possibly a grinder).

It has been said that the most common size drill is a No. 30 with the $^1/_8$" drill close behind. Only one percent (1%) of all the drills sold are over $^3/_4$", ninety percent (90%) of the drills sold fall between 0.05" and 0.400" in size.

Constructing Tool Numbers and Aux. I.D.'s for Drills

DRILL TYPE	SIZE	MODIFIER	SPECIAL CONDITION
		HIGH SPIRAL	
		LOW SPIRAL	
		HEAVY DUTY	
		LEFT HAND	
		MILLIMETER	
		TAPER LENGTH	
		-HIGH SPIRAL	XTRA LONG
		TAPER SHANK	BRIGHT FINISH
	FRACTIONAL	-DIFF. TAPER SIZE	SURFACE TREAT
	WIRE GAGE NO.	-CHIPBREAKER	COBALT
	LETTER	-HEAVY DUTY	CARBIDE
		-COBALT	CARBIDE TIPPED
		STUB DRILL	
TWIST		AIRCRAFT LENGTHS	
DRILLS		AUTOMOTIVE	
		-TAPER LENGTH	
		OIL HOLE	
		MIST COOLANT	
		STRAIGHT SHANK	
	PLAIN		
	BELL		
CENTER	RADIAL		
DRILLS	HEAVY DUTY	SIZE NO.	
	SPOTTING		
	TIPS		
	-HEAVY DUTY		
	-RADIAL		
HALF	FRACTIONAL	RIGHT HAND	
ROUND	JOBBERS		
	WIRE GAGE NO.	LEFT HAND	
SILVER			
DEMING	FRACTIONAL	SHANK SIZE-1 /2 STD.	
COUNTER			
DRILLS	SCREW SIZE		
CORE	FRACTIONAL	3 FLUTE	STRAIGHT SHANK
DRILLS		4 FLUTE	TAPER SHANK
SPADE	CUTTING DIA.	STANDARD	
DRILLS		CORE	
SPADE		STRAIGHT SHANK	
DRILL	SERIES	TAPER SHANK	COOLANT
HOLDERS		XTRA LONG	

TWIST, CENTER, HALF ROUND, SILVER-DEMING, COUNTER, CORE & SPADE DRILLS

Tool Number Begins	General Description	TLC Aux. I.D.
TD	Twist Drills - Fractional, Wire Gage No., Letter, Millimeter - Stub Length, Aircraft, Automotive, Oil Hole, Mist Coolant - Straight & Taper Shank, Taper Length, General Purpose, Right & Left Hand, Low & High Spiral, Heavy Duty, Cobalt, Chipbreaker	TD
TDC	Core Drills	TDC
TDSD	Silver Deming Drills	TDSD
CD	Center Drills	CD
CSD	Counter Drills	CSD
HRD	Half Round Drills	HRD
SD	Spade Drills	SD
SDH	Spade Drill Holders	SDH

Notes:

1) Cobalt, heavy duty, and cotter pin drills generally have 135 degree tool point angles and are heavier in construction than general purpose or jobber's drills. Special application drills may need additional modifiers for your drills, such as double/triple margins, step diameters, and drill/reamer combinations. 2) Generally drills under $^{1}/_{16}$" are not split points. 3) Taper lengths come standard in fractional and wire gage no. sizes only with high spiral flutes as an alternate. 4) Screw machine drills come standard in fractional, wire gage no., and letter sizes with left hand spiral as an alternate. Fractional diameters under 1-$^{1}/_{16}$" have shank sizes equal to diameter, sizes through 1-$^{1}/_{4}$" have 1" shanks, through 1-$^{1}/_{2}$" have 1-$^{1}/_{4}$" shanks, sizes 1-$^{9}/_{16}$" and over have 1-$^{1}/_{2}$" shanks. All are standard 118 degree. points. 5) Flute lengths of wire gage no. sizes 81 through 90 are approximately 8 times the diameter. 6) Aircraft length drills of 6" overall length (OAL) have 118 degree points, and 12" OAL have 135 degree split points. Wire gage sizes 53 through 60 are not split points.

TOOL NUMBER CONSTRUCTION GUIDE — *DRILLS*

TWIST DRILLS, JOBBERS/GENERAL PURPOSE, TAPER LENGTH, TAPER SHANK, SCREW MACHINE, AIRCRAFT EXTENSION, AUTOMOTIVE, MIST COOLANT, OIL HOLE, SILVER-DEMING, CORE DRILLS

TOOL NO.: Class Code/TYPE/FRACTION, etc./MODIFIER/*SPEC
Core Drills TOOL NO.: TD/#FLUTES/C/FRACTION/MODIFIER

CLASS	TYPE	SIZE	MODIFIER	SPECIAL
TD	**TS**	**1-11/16**	**HD**	**S**
TWIST	TL Taper Length	FRACTIONS 1-11/16	LS Low Spiral	B Bright Finish
DRILL	AI Aircraft Length	WIRE GAGE NO. #61	HS High Spiral	S Surface Treated
	AU Automotive	LETTER A	HD Heavy Duty	L Long Length
	MC Mist Coolant	MILLIMETER 9MM	LH Left Hand	X tra Long
	OH Oil Hole	DECIMAL 1.6875	MM Millimeter	SS Straight Shank
	SD Silver Deming		CO Cobalt	TS Taper Shank
	TS Taper Shank		CB Chipbreaker	TL Taper Length
	xC* Core Drill			-# Shank Deviation
	S Stub Drill			G Gold Oxide
				CT Carb.Tipped

* *Note*: In xC, "x" equals the number of flutes.

TD CLASS OF TOOLS: TOOL DESCRIPTION FORMAT	
TD:	DIA.,FLUTELNGTH,OAL,SHANK,SPEC.,TWISTDRILL
TDTL:	DIA.,FLUTELNGTH,OAL,TAPERLENGTH,(HS), TWISTDRILL
TDS:	DIA.,FLUTELNGTH,OAL,SCREWMACH(LFTHND)TWISTDRILL
TDAI:	DIA.,FLUTELNGTH,OAL,AIRCRFTEXT,TWISTDRILL
TDAU:	DIA.,FLUTELNGTH,OAL,AUTOSSTANGD(TL),TWSTDRILL
TDMC:	DIA.,FLUTELNGTH,OAL,TAPERNO.,MISTCOOLNT DRILL
TDOH:	DIA.,FLUTELNGTH,OAL,THRDHOLESIZE,OILHOLE,TWISTDRILL
TDSD:	DIA.,FLUTELNGTH,OAL,SILVERDEMING,TWISTDRILL
TDTS:	DIA.,FLUTELNGTH,OAL,SHNKSZE,MODIF.,TAPRSHNK,TWISTDRLL
TDxC:	DIA.,FLUTELNGTH,OAL,NO.FLUTES,SHANKSIZE,COREDRILL

Aux. Code Format and Dimension Definition
Aux. I.D.: [TD, TDTL, TDS, TDAI, TDAU, TDMC, TDOH, TDSD, TDTS, TDC] D/1: Diameter D/2: Flute Length

EXAMPLES: TWIST DRILLS

Tool No.	Description	Aux.	D/1	D/2
TD1/8	1/8D,1-5/8FL,2-3/4OAL,SSJB,TWISTDRL	TD	0.125	1.625
TD1/8LS	1/8D,1-5/8FL,2-3/4OAL,SSJBLS,TWISTD	TD	0.125	1.625
TD1/8HD	1/8D,1-5/8FL,2-3/4OAL,SSJBHD,TWISTD	TD	0.125	1.625
TD1/8HS	1/8D,1-5/8FL,2-3/4OAL,SSJBHS,TWISTD	TD	0.125	1.625
TDTL1-15/32	1-15/32D,9-1/2FL,14-7/8OAL,TL,TWSTD	TDTL	1.9375	9.5000
TDTL1/2HS	1/2D,4-3/4FL,7-3/4OAL,TLHS,TWSTDRLL	TDTL	0.5000	4.7500
TDS1-15/32	1-15/32D,5-3/8FL,8-1/2OAL,SCRWMCHTD	TDS	1.9375	5.3750
TDS1/2LH	1/2D,2-1/4FL,3-3/4OAL,SCRWMCHLH,TWD	TDS	0.5000	2.2500
TDAI1/4	1/4D,2-3/4FL,6OAL,AIRCRFTEXT,TWSTDR	TDAI	0.250	2.750
TDAI1/4L	1/4D,2-3/4FL,12OAL,AIRCRFTEXT,TWSTD	TDAI	0.250	2.750
TDAU9MM	.3543D,3-1/2FL,4-7/8OAL,AUTOSS,TWST	TDAU	0.3543	3.500
TDAU23/64	23/64D,3-1/2FL,4-7/8OAL,AUTOSS,TWST	TDAU	0.3594	3.500
TDAU23/64TL	23/64D,4-1/4FL,6-3/4OAL,AUTOSSTL,TD	TDAU	0.3594	4.250
TDAUU	.368D,3-5/8FL,5OAL,AUTOSS,TWSTDRLL	TDAU	0.3680	5.000
TDMC1-7/16	1-7/16D,9-1/8FL,14-3/4OAL,#4TS,MIST	TDMC	1.4375	9.125
TDOH1-7/16	1-7/16D,9-5/8FL,14-3/4OAL,3/8THD,OH	TDOH	1.4375	9.625
TDSD1-3/16	1-3/16D,3-1/8FL,6OAL,SLVRDMNG,TWSTD	TDSD	1.1875	3.125
TDTS1-11/16CB	1-11/16D,10-1/8FL,17-1/8OAL,#5TS,CB	TDTS	1.6875	10.125
TDTS1-11/16	1-11/16D,10-1/8FL,17-1/8OAL,#5TS,TD	TDTS	1.6875	10.125
TDTS1-11/16-4	1-11/16D,10-1/8FL,15-3/4OAL,#4TS,CB	TDTS	1.6875	10.125
TDTS1-3/32HD	1-3/32D,6-7/8FL,12-1/2OAL,#4TS,HDTD	TDTS	1.0938	6.875
TDTSW	.386D,3-5/8FL,7OAL,#1TS,TPRSHNK,TD	TDTS	0.386	3.625
TD3C15/16	15/16D,6-1/8FL,10-3/4OAL,#3TS,CORE	TDC	0.9375	6.125
TD4C15/16SS	15/16D,6-1/8FL,10-3/4OAL,SS,COREDR	TDC	0.9375	6.125

TOOL NUMBER CONSTRUCTION GUIDE — CENTER DRILLS

CD — CENTER DRILLS (ALL TYPES)

TOOL NO.: CD/TYPE/SIZE

CLASS	TYPE	SIZE	MODIFIER
CD	**R**	**00**	**R**
CD CENTER DRILL	R Radial	SERIES # 00	R Radial
	H Heavy Duty	FRACTIONAL $^3/_4$	S Heavy Duty
	S Spotting		R & S basically for TIPS
	T Tips		
	P Plain		
	B Bell (omit if standard)		

CD CLASS OF TOOLS : TOOL DESCRIPTION FORMAT
CD : SIZE,DIA,BODYDIA,OAL — for R OR H TYPES
CD : DIA,FLUTELNGTH,OAL — for STD. OR S TYPES
CD : SIZE,DRILLDIA,BODYDIA,OAL — for P OR B TYPES
CD : SIZE,BODY x POINT DIA — for T TYPE

Aux. Code Format and Dimension Definition

Aux. I.D.: CD D/1: Diameter/Size D/2: Flute Length

EXAMPLES: CD CLASS CENTER DRILLs

Tool No.	Description	Aux.	D/1	D/2
CDR00	00,.025D,5/64BODY,7/8OAL,RADIAL,CNTR	CD	00	
CDH18	18,1/4D,3/4BODY,3-1/2OAL,HVYDTY,CNTR	CD	18	
CD11/64	11/64D,1FL,1-1/2OAL,CENTERDRILL	CD	0.1719	1.0000
CDS3/4	3/4D,1-1/8FL,2-1/4OAL,SPOTTINGDRILL	CD	0.750	1.1250
CDP1	1,3/64D,1/8BODY,1-1/4OAL,PLAIN,CNTR	CD	1	
CDB18	18,1/4D,3/4BODY,3-1/2OAL,BELL,CNTRDR	CD	18	
CDT0R	0R,1/8x.031,RADIAL,CNTR DRL TIP	CD	0.031	
CDT0S	0S,1/8x.031,HVYDTY,CNTR DRL TIP	CD	0.031	

TOOL NUMBER CONSTRUCTION GUIDE — COUNTERSINK DRILLS

CSD — COUNTERSINK DRILLS, FLAT & OVAL HEAD SCREWS, 82 DEG INC. ANG.

TOOL NO.: CSD/SCREW SIZE

CLASS	SIZE
CSD COUNTER SINK DRILL	**#12** SCREW SIZE

CSD Class of Tools : Tool Description Format
CSD: SCREWSIZE,DIA,BODYDIA,SHANKDIA,OAL,C'SINK DRILL

Aux. Code Format and Dimension Definition
Aux. I.D.: CSD D/1: Screw Size D/2: Drill Diameter

EXAMPLE: CSD CLASS COUNTERSINK DRILLS

Tool No.	Description	Aux.	D/1	D/2
CSD#12	#12,.22D,7/16BODY,1/4SHNK,2OAL,CSNK	CSD	12	0.220

TOOL NUMBER CONSTRUCTION GUIDE — HALF ROUND DRILLS

HRD — HALF ROUND STRAIGHT SHANK JOBBERS LENGTH DRILL

TOOL NO.: HRD/FRACTION, etc./MODIFIER

CLASS	SIZE	MODIFIER
HRD	**15/32**	**L**
HALF	FRACTIONAL	L
ROUND	15/32	Left Hand
DRILL		
	WIRE GAGE NO.	
	#10	

HRD CLASS OF TOOLS: TOOL DESCRIPTION FORMAT
HRD: DIA, FLUTELNGTH, OAL, HALFROUND SS DRILL

Aux. Code Format and Dimension Definition
Aux. I.D.: HRD D/1: Diameter D/2: Flute Length

EXAMPLE: HRD CLASS HALF ROUND DRILLS

Tool No.	Description	Aux.	D/1	D/2
HRD15/32	15/32D,4-5/16FL,5-3/4OAL,1/2RNDSS	HRD	0.4688	4.3125

TOOL NUMBER CONSTRUCTION GUIDE — SPADE DRILL BLADES

SD & SDC — SPADE DRILL BLADES, M3 TYPE 2, STANDARD & CORE BLADES

TOOL NO.: SD OR SDC/CUTTING DIA

CLASS	TYPE	SIZE
SD	**C**	**1-15/16**
SPADE	C	FRACTIONAL
DRILL	Core	1-15/16
BLADE		
	(omit if std.)	

SD & SDC CLASS OF TOOLS: TOOL DESCRIPTION FORMAT
SD or SDC: CUTTING DIA, SERIES, STD/CORE, SPADE DRILL BLADE

Aux. Code Format and Dimension Definition
Aux. I.D.: SD OR SDC D/1: Diameter D/2: Maximum Cut

EXAMPLES: SD CLASS SPADE DRILL BLADES

Tool No.	Description	Aux.	D/1	D/2
SD1-15/16	1-15/16D,C,STD,SPADE DRILL BLADE	SD	1.9375	1.9375
SDC1-15/16	1-15/16D,C,CORE,SPADE DRILL BLADE	SDC	1.9375	0.3125

TOOL NUMBER CONSTRUCTION GUIDE — SPADE DRILL BLADE HOLDERS

SDH — SPADE DRILL BLADE HOLDER

TOOL NO.: SDH/SERIES/SHANKSIZE/MODIFIER

CLASS	SERIES	SHANK	MODIFIER
SDH	**AA**	**1**	**L**
SPADE	SERIES OF BLADE	FRACTIONAL	C
DRILL	AA	1	Coolant Hole
BLADE			
HOLDER		MORSE TAPER#	L
		3	Long
			X
			Xtra Long

SDH CLASS OF TOOLS: TOOL DESCRIPTION FORMAT
SDH: SERIES, RANGE, SHANK, MODIFIER, SPADE DRILL BLADE HOLDER

Aux. Code Format and Dimension Definition
Aux. I.D.: SDH D/1: Minimum. Range D/2: Maximum Range

EXAMPLES: SDH CLASS SPADE DRILL HOLDERS

Tool No.	Description	Aux.	D/1	D/2
SDHAA1	AA,1-1/32*1-1/4,1DSHNK,SPDRBLHLDR	SDH	1.031	1.250
SDHAA#3	AA,1-1/32*1-1/4,#3MT,SPDRBLHLDR	SDH	1.031	1.250
SDHAA1.5	AA,1-1/32*1-1/4,1.5SHNK,SPDRBLHLDR	SDH	1.031	1.250
SDHAA1L	AA,1-1/32*1-1/4,1DSHNK,LNG,SPDRHLDR	SDH	1.031	1.250
SDHB#4CL	B,1-9/32*1-1/2,#4MT,CL,SPDRBLHLDR	SDH	1.281	1.500
SDHB#4CX	B,1-9/32*1-1/2,#4MT,CX,SPDRBLHLDR	SDH	1.281	1.500

II. Reamers

Drills are used to make holes, and reamers are used to finish holes to size. Selection of the right reamer requires consideration of many factors, such as the piece part material, diameter, length of hole, depth of hole, condition of the hole (i.e., blind, through, or interrupted), number of holes, and cost of tooling. The reamer selection is based on considering the various tool materials that reamers come in, the right geometry for the job, the flute design, and the surface treatment, if any, needed. After looking at all these conditions and requirements, the machine, its spindle, and the workpiece itself offer conditions that merit consideration. Selecting the right reamer to do the job is not always straightforward.

The basic tool material, HSS, comprises the majority of reamers, and is suitable for most jobs. Other popular tool materials are solid carbide, carbide tipped, and cobalt-HSS. Selection of the reamer's tool material is often a practical decision based on tool costs and the size of the production run.

Some interesting basics about reamers you may wish to reacquaint yourself with are the following.

- Use carbide where heat and abrasion are a problem. Soft materials like diecast aluminum or cast iron can be abrasive, and materials above Rc 40 can be more than HSS can handle.
- The machine should be in good condition; loose or misaligned spindles will rob performance and damage both the reamer and piece part.
- First rule — start with a straight flute reamer.
- Second rule — if it chatters, chips won't clear, or the wall of the hole becomes scored, try a left-hand spiral/right-hand cut or a right-hand spiral/right-hand cut.
- Above $^{3}/_{4}$" diameters, consider shell reamers.
- Expansion reamers on long production runs can be more cost effective.
- Reamers size holes, they don't straighten them.
- Adjustable reamers are generally selected when the hole is an odd size.

Constructing Tool Numbers and Aux. I.D.'s for REAMERS

Considerations Used for General Tool Number Construction

Class	Size	Primary Modifier	Secondary Modifier
		STRAIGHT FLUTE	
TYPE		RIGHT HAND SPIRAL	
OF		LEFT HAND SPIRAL	
REAMER	FRACTIONAL	LEFT HAND CUT	HSS
	WIRE GAGE NO.	RIGHT HAND CUT	COBALT
	LETTER	HELICAL FLUTE	CARBIDE
	DOWEL PIN	TAPERED	
		TAPER SHANK	
CHUCKING		-B&S	
HAND		-MORSE	
ROSE		STRAIGHT SHANK	
TAPER PIN			
PIPE TAPER	SQUARE END	RIGHT HAND CUT	LEFT HAND SPIRAL
SCREW MACH	STRAIGHTSHANK	RIGHT HAND CUT	LEFT HAND SPIRAL
DIE MAKERS	HELICAL FLUTE	STRAIGHT SHANK	
		RIGHT HAND SPIRAL	ARBOR
SHELL	FRACTIONAL	RIGHT HAND CUT	-TAPER SHANK
		LEFT HAND SPIRAL	-STRAIGHT SHANK
		STRAIGHT FLUTE	
EXPANSION			STRAIGHT SHANK
CHUCKING	FRACTIONAL	STRAIGHT FLUTE	TAPER SHANK
B&S TAPER	TAPER SIZE	STRAIGHT SHANK	STRAIGHT FLUTE
MORSE			
TAPER	TAPER SIZE	SQUARE SHANK	
		TAPER SHANK	
	BORE RANGE	PILOT	
		BLADES	

REAMERS

Tool Number Begins	General Description	TLC Aux. I.D.
CR	Chucking Reamers - Fractional, Wire Gage No., Letter	RC
CRD	Dowel Pin Reamers	RDC
CRT	Taper Shank Chucking Reamers	RCT
CRTJ	Taper Shank Jobbers Reamers	RCTJ
CRR	Rose Chucking Reamers	RCR
SR	Shell Reamers	RS
SRA	Shell Reamer Arbors	RSAS - Straight RSAT - Taper
CRE	Expansion Chucking Reamers	RECS - Straight RECT - Taper
BSTR	B&S Taper Reamer	RBST
MTR	Morse Taper Reamer	RMTS - Straight RMTT - Taper
AR	Adjustable Blade Reamer	RA
HR	Hand Reamers	RH
TPTR	Taper Pipe Reamer	RTPT
RTP	Taper Pin Reamer	RTP
SMR	Stub Screw Machine Reamers	RSM
DMR	Die Maker Reamers	RDM

TOOL NUMBER CONSTRUCTION GUIDE — REAMERS

CR – CHUCKING REAMERS, **CRD** – DOWEL PIN, **CRT** – TAPER SHANK, **CRTJ** – JOBBERS REAMERS, **CRR** – ROSE, **HR** – HAND, **TPTR** – TAPER PIPE, **TPR** – TAPER PIN, **SR** – SHELL, **SRA** – SHELL REAMER ARBORS

TOOL NUMBER: Class Code/SIZE/MODIFIER
SMR – STUB SCREW MACHINE – **TOOL NO.: SMR/SERIES**
DMR – DIE MAKER – **TOOL NO.: DMR/LETTER**
CRE – EXPANSION CHUCKING – **TOOL NO.: CRE/DIA/MODIFIER**
BSTR – B&S TAPER – **TOOL NO.: BSTR/TAPERNO**.
MTR – MORSE TAPER – **TOOL NO.: MTR/TAPERNO./MODIFIER**
AR – ADJUSTABLE REAMER – **TOOL NO.: AR/MEDIAN RANGE**

CLASS	SIZE	MODIFIER
CR	**3/64**	**RSC**
CR	FRACTIONAL	RSC
Chucking Reamers	1-11/16	Right Hand Spiral
CRD		Right Hand Cut
Dowel Pin	WIRE GAGE NO.	LSRC
CRT	#61	Left Hand Spiral
Taper Shank		Right Hand Cut
CRTJ	LETTER	LS
Taper Shank	A	Left Hand Spiral
Jobbers		
CRR	DECIMAL	SF
Rose Style	1.1240	Straight Flute
CRE		
Expansion Style	SIZE NO.	HF
AR	7/0	Helical Flute
Adjustable Reamer		
HR	SERIES	S
Hand Reamer	23	Straight Shank
TPTRS		
Taper Pipe Thread	TAPER NO.	T
TPR	1	Taper Shank
Taper Pin Reamer		
SMR	MEDIAN RANGE	SQ
Stub Screw Mach.	0.2969	Square Shank
DMR		
Die Makers		
SR		
Shell Reamer		
BSTR		
B&S Taper		
MTR		
Morse Taper		
SRA		
Shell Reamer		
Arbors		

REAMER CLASS OF TOOLS: TOOL DESCRIPTION FORMATS
CR,CRD,CRT,CRTJ,CRR,HR,CRE: DIA,NO.FLUTES,FLUTELNGTH,OAL,MOD.S

Aux. Code Format and Dimension Definition
Aux. I.D.: [RC, RCD, RCT, RCTJ, RCR, RH, RCE]
D/1: Diameter D/2: Flute Length

REAMER CLASS OF TOOLS:	TOOL DESCRIPTION FORMAT
TPTR:	SIZE,FLNGTH,OAL,SHNKDIA,TAPERPIPEREAMER
TPR:	SIZE,FLNGTH,OAL,MOD,TAPERPINREAMER

Aux. Code Format and Dimension Definition
Aux. I.D.: [RTPT, RTP] D/1: Size D/2: Flute Length

REAMER CLASS OF TOOLS:	TOOL DESCRIPTION FORMAT
SMR:	SERIES,FLNGTH,OAL,RANGE,SCREWMACHREAMER

Aux. Code Format and Dimension Definition
Aux. I.D.: RSM D/1: Minimum Range D/2: Maximum Range

REAMER CLASS OF TOOLS:	TOOL DESCRIPTION FORMAT
DMR:	LETTER,NO.FLUTES,FLNGTH,OAL DIE REAMER

Aux. Code Format and Dimension Definition
Aux. I.D.: RDM D/1: Small Diameter D/2: Large Diameter

REAMER CLASS OF TOOLS:	TOOL DESCRIPTION FORMAT
SR:	DIA.,NO.FLUTES,,OAL,MOD.,ARBORNO.,SHELLREAMER

Aux. Code Format and Dimension Definition
Aux. I.D.: RS D/1: Diameter D/2:

REAMER CLASS OF TOOLS:	TOOL DESCRIPTION FORMAT
SRA:	SIZE,SHANKDIA/MORSETAPERNO.,OAL,SHELLRMER ARBOR

Aux. Code Format and Dimension Definition
Aux. I.D.: RSA D/1: Size Number D/2:

REAMER CLASS OF TOOLS:	TOOL DESCRIPTION FORMAT
BSTR - MTR:	TAPERNO.,FLNGTH,OAL,SHNKDIA/TAPER

Aux. Code Format and Dimension Definition
Aux. I.D.: [RBST,RMT] D/1: Taper Number D/2: Flute Length

REAMER CLASS OF TOOLS:	TOOL DESCRIPTION FORMAT
AR:	RANGE,NO.BLADES,BLADELNGTH,OAL,ADJSTBLE REAMER

Aux. Code Format and Dimension Definition
Aux. I.D.: RA D/1:Low Range D/2: High Range

EXAMPLES: REAMERS

Tool No.	Description	Aux.	D/1	D/2
CR3/64	3/64D,4F,1/2FL,2-1/2OAL,SS,SFL,REAM	RC	0.0469	0.5000
CR3/64RSC	3/64D,4F,1/2FL,2-1/2OAL,SS,RSC,REAM	RC	0.0469	0.5000
CR#1	.228D,6F,1-1/2FL,6OAL,SS,SFL,REAMER	RC	0.2280	1.500
CR.1240	.124UD,4F,7/8FL,3-1/3OAL,SS,SFL,REA	RC	0.1240	0.8750
CRE	.25D,6F,1-1/2FL,6OAL,SS,SFL,REAMER	RC	0.125	1.500
CRD.1230	.1230D,4F,7/8FL,3-1/2OAL,SS,SFL,REA	RCD	0.1230	0.8750
CRT1/4	1/4D,6F,1-1/2FL,6OAL,#1MT,SFL,REAMR	RCT	0.2500	1.500
CRT1/4RSC	1/4D,6F,1-1/2FL,6OAL,#1MT,RSC,REAMR	RCT	0.2500	1.500
CRTJ1/4	1/4D,6F,2FL,5-3/16OAL,#1MT,JOBBERRM	RCTJ	0.2509	2.000
CRR1/16	.0625D,4F,1/2FL,2-1/2OAL,SS,SFLREAM	RCR	0.0625	0.5000
CRR1/16RSC	.0625D,4F,1/2FL,2-1/2OAL,SS,RSCREAM	RCR	0.0625	0.5000
HR1/4	1/4D,6F,2FL,4OAL,SS,SFL,HAND REAMER	RH	0.2500	2.000
HR1/4LSRC	1/4D,6F,2FL,4OAL,SS,LSRC,HAND REAMER	RHC	0.2500	2.000
TPTR1-1/4	1-1/4,1-3/4FL,4OAL,1-5/16SHNK,PIPE	RTPT	1.250	1.750
TPR7/0LS	7/0,13/16FL,1-13/16OAL,LS,TAPERPIN	RTP	7.0	0.8125
TPR7/0SF	7/0,13/16FL,1-13/16OAL,SF,TAPERPIN	RTP	7.0	0.8125
TPR7/0HF	7/0,13/16FL,1-13/16OAL,HF,TAPERPIN	RTP	7.0	0.8125
SMR23	23,1-1/2FL,3OAL,.9421/1.010,SCRMCRM	RSM	0.9421	1.010
DMRAAA	AAA,2F,1-1/8FL,2-1/4OAL,DIE REAMER	RDM	0.0550	0.0700
SR3/4SF	3/4D,10F,2-1/4OAL,SF,#4MTSHK,SHELLR	RS	0.7500	0.000
SR3/4LS	3/4D,10F,2-1/4OAL,LS,#4MTSHK,SHELLR	RS	0.7500	0.000
SRA4T	4,#2MT,9OAL,SHELLREAMERARBOR	SRA	4.000	0.000
ECR7/16S	7/16D,6F,7/8FL,7OAL,SS,EXPANREAMER	RCE	0.4375	0.8750
ECR7/16T	7/16D,6F,7/8FL,7OAL,#1MT,EXPANREAM	RCE	0.4375	0.8750
BSTR1	1,2-7/8FL,4-3/4OAL,9/32DSHNK,B&STPR	RBST	1.000	2.875
MTR0SQ	0,2-1/4FL,3-3/4OAL,5/16DSHNK,MTREAM	RMT	0.000	2.250
MTR0T	0,2-1/4FL,5-11/32OAL,0TSHNK,MTREAM	RMT	0.000	2.250
AR.2969	.281/.312D,4B,1-5/8BL,3-3/4OAL,ADJRM	RA	0.2813	0.3125

III. Taps and Dies

Taps and dies, like other tools, are made from a variety of tool materials. Again, abrasion, temperature, and piece part material influence the tap or die material selected, as well as the class of fit and size required. Other influences of selection (predominately taps) and some recommendations areas follow.

- Spiral pointed taps are a good choice for through holes; or straight flute taps of 2, 3, or 4 flutes are suitable, depending on size.
- Use bottoming taps for blind holes, and a spiral fluted style for interrupted holes.
- Try to tap no deeper than 1.5 times the diameter.
- Use taper taps to reduce the chip load per tooth.
- Use high spirals on very ductile materials.
- Use raking and hooking to control chip breaking and chip flow.
- Xrelief or Zero flute taps aren't used often enough. They have good rigidity, finish, tap well at angles, blind holes, work well in ductile metals or below Rc35.
- Use surface treated taps to solve problems.
- The wrong surface treatment will create problems.
- Remember, taps are for holes, and dies are for external threads.

Constructing Tool Numbers and Aux. I.D.'s for TAPS

Considerations Used for General Tool Number Construction

TOOL TYPE	STYLE	MODIFIER	MODIFIER
TAPS	PLUG	STRAIGHT FLUTES	SPIRAL POINT
	BOTTOM	STUB FLUTES	HIHOOK
FRACTIONAL	TAPER	SPIRAL FLUTES	NEUTRAL HOOK
SCREW NO.		HISPIRAL FLUTES	INTERRUPTED
METRIC	2 FLUTE		LEFT HAND
PIPE	3 FLUTE		
	4 FLUTE		
PULLEY	6 FLUTE		
NUT			
	L1,H1 THRU H11		
STD. HAND			
	0 FLUTE		
	XRELIEVED		

TAPS

Tool Number Begins	Description	TLC Aux. Code
(No Entry)	Standard Taps	TAP
P	Pipe Taps	TAPP
A	Acme Taps	TAPA
H	High Spiral Taps	TAPH
S	Stub Flute Taps	TAPS
T	Tapper Taps	TAPT
N	Nut Taps	TAPN
Y	Pulley Taps	TAPY
	Dies - Adjustable	DIEA
	Rethreading (Repair)	DIER

Notes:

1) All threads are assumed to be classified as ground. 2) Tolerances are specified by the standards tables according to the L and H number defined limits. 3) UNC — NC and UNF - NF designations are not included in tool numbers. 4) Refer to your normal recommendations chart for tolerance limits (H class numbers) for class 2, 3, 2B, 3B threads.

TOOL NUMBER CONSTRUCTION GUIDE — TAPS

TAPS – FRACTIONAL, MACHINE SCREW, PIPE, METRIC

TOOL NO.: Class Code/TAPSIZE-TPI/NO.FLUTE/P-B-T/HTOL./MODIFIER

CLASS	SIZE	# FLUTES	STYLE	TOLER.	MODIFIER
	¼-20	**2F**	**P**	**H3**	**L**
Omit	Tap Size	# FLUTES	P	L1	L
if	-	0	Plug	H1	Left Hand
Std.	Threads/In.	2		through	
	¼-20	3	B	H11	S
S	0-80	4	Bottom		Spiral Point
Stub		6		P1	
Flute	Omit for		T	through	H
	Pipe Taps		Taper	P8	High Hook
A					
Acme		NPS		B1	N
		NPSF		through	Neutral Hook
P		NPT		B8	
Pulley		NPTF			I
					Interrupted
T					
Tapper					LS
					Long Shank
N					or Large
Nut					
					H
T					High
Tandem					Spiral
					SF
					Spiral Flute

TAPS CLASS OF TOOLS: TOOL DESCRIPTION FORMAT
TAP: : NO. FLUTES,THRDLNGTH,OAL,P-B-T,HTOL,MODIFIER,TAP

Aux. Code Format and Dimension Definition

Aux. I.D.: TAPx D/1: Size D/2: Threads per Inch

x = CLASS of TAP

EXAMPLES: TAPS – Standard Hand Tap Straight Flutes

Tool No.	Description	Aux	D/1	D/2
1/4-202FPH3	2F,1TL,2-1/2OAL,PLUG,H3,STF,HANDTAP	TAP	0.25	20
1/4-200FPP4	0F,1TL,2-1/2OAL,PLUG,P4,0FLUTE TAP	TAP	0.25	20
1/4-202FPH3L	2F,1TL,2-1/2OAL,PLUG,H3,STF,LHNDTAP	TAP	0.25	20
1/4-202FBH3	2F,1TL,2-1/2OAL,BOTT,H3,STF,HANDTAP	TAP	0.25	20
1/4-200FBB4	0F,1TL,2-1/2OAL,BOTT,B4,0FLUTE TAP	TAP	0.25	20
1/4-203FPH3	3F,1TL,2-1/2OAL,PLUG,H3,STF,HANDTAP	TAP	0.25	20
1/4-203FPH1	3F,1TL,2-1/2OAL,PLUG,H1,STF,HANDTAP	TAP	0.25	20
1/4-203FBH5	3F,1TL,2-1/2OAL,BOTT,H5,STF,HANDTAP	TAP	0.25	20
1/4-204FTH1	4F,1TL,2-1/2OAL,TPER,H1,STF,HANDTAP	TAP	0.25	20
1/4-204FPH3	4F,1TL,2-1/2OAL,PLUG,H3,STF,HANDTAP	TAP	0.25	20
1/4-204FBH3	4F,1TL,2-1/2OAL,BOTT,H3,STF,HANDTAP	TAP	0.25	20
1-1/2126FTH4	6F,3TL,6-3/8OAL,TPER,H4,STF,HANDTAP	TAP	1.5	12
1-1/2126FPH4	6F,3TL,6-3/8OAL,PLUG,H4,STF,HANDTAP	TAP	1.5	12
1-1/2126FPH4L	6F,3TL,6-3/8OAL,PLUG,H4,STF,LHNDTAP	TAP	1.5	12
1-1/2126FBH4	6F,3TL,6-3/8OAL,BOTT,H4,STF,HANDTAP	TAP	1.5	12
1-1/264FTH4	4F,3TL,6-3/8OAL,TPER,H4,STF,HANDTAP	TAP	1.5	6

EXAMPLES: TAPS – Standard Hand Tap – Screw Machine Taps Straight Flutes

Tool No.	Description	Aux.	D/1	D/2
0-802FTH1	2F,5/16TL,1-5/8OAL,TPER,H1,STF,SCREW	TAP	0	80
0-802FPH1	2F,5/16TL,1-5/8OAL,PLUG,H1,STF,SCREW	TAP	0	80
0-802FPH1L	2F,5/16TL,1-5/8OAL,PLUG,H1,STFLH,SCR	TAP	0	80
10-322FPH1	2F,7/8TL,2-3/8OAL,PLUG,H1,STF,SCREW	TAP	10	32
0-802FBH1	2F,5/16TL,1-5/8OAL,BOTT,H1,STF,SCREW	TAP	0	80
0-800FBB2	0F,5/16TL,1-5/8OAL,BOTT,B2,0FL,SCREW	TAP	0	80
2-563FTH2	3F,7/16TL,1-3/4OAL,TPER,H2,STF,SCREW	TAP	2	56
2-563FPH2	3F,7/16TL,1-3/4OAL,PLUG,H2,STF,SCREW	TAP	2	56
10-243FPH3	3F,7/8TL,2-3/8OAL,PLUG,H3,STF,SCREW	TAP	10	24
2-563FBH2	3F,7/16TL,1-3/4OAL,BOTT,H2,STF,SCREW	TAP	2	56
10-323FBH2	3F,7/8TL,2-3/8OAL,BOTT,H2,STF,SCREW	TAP	10	32
8-324FTH3	4F,3/4TL,2-1/8OAL,TPER,H3,STF,SCREW	TAP	8	32
12-284FTH3	4F,15/16TL,2-3/8OAL,TPER,H3,STF,SCRE	TAP	12	28
8-324FPH3	4F,3/4TL,2-1/8OAL,PLUG,H3,STF,SCREW	TAP	8	32
12-284FPH3	4F,15/16TL,2-3/8OAL,PLUG,H3,STF,SCRW	TAP	12	28
12-280FPP4	0F,15/16TL,2-3/8OAL,PLUG,P4,0FL,SCRW	TAP	12	28
12-284FPH3L	4F,15/16TL,2-3/8OAL,PLUG,H3,STFLH,SC	TAP	12	28
8-324FBH3	4F,3/4TL,2-1/8OAL,BOTT,H3,STF,SCREW	TAP	8	32
12-284FBH3	4F,15/16TL,2-3/8OAL,BOTT,H3,STF,SCRW	TAP	12	28

EXAMPLES: TAPS – Machine Screw 0.002 over H3 limits = H7

Tool No.	Description	Aux.	D/1	D/2
6-323FPH7	3F,11/16TL,2OAL,PLUG,H7,STF,SCREW	TAP	6	32
10-324FPH7	4F,7/8TL,2-3/8OAL,PLUG,H7,STF,SCRE	TAP	10	32
6-323FBH7	3F,11/16TL,2OAL,BOTT,H7,STF,SCREW	TAP	6	32
10-324FBH7	4F,7/8TL,2-3/8OAL,BOTT,H7,STF,SCRE	TAP	10	32

TAP MODIFIER S-SPIRAL POINT
EXAMPLES: Standard Hand Tap Straight Flutes Spiral Point

Tool No.	Description	Aux.	D/1	D/2
1/4-202FPH3S	2F,1TL,2-1/2OAL,PLUG,H3,STF,SPT,TAP	TAP	0.25	20
5/16-242FPH3	2F,1-1/8TL,2-23/32OL,PLG,H3,STF,SPT	TAP	0.3125	24
1/4-202FBH3S	2F,1TL,2-1/2OAL,BOTT,H3,STF,SPT,TAP	TAP	0.25	20
5/16-242FBH3S	2F,1-1/8TL,2-23/32OL,BTT,H3,STF,SPT	TAP	0.3125	24
3/8-163FPH3S	3F,1-1/4TL,2-15/16OL,PLG,H3,STF,SPT	TAP	0.375	16
5/8-113FPH3S	3F,1.812TL,3.812OAL,PLUG,H3,STF,SPT	TAP	0.625	11

EXAMPLES: Standard Hand Tap Screw Machine Taps Straight Flutes Spiral Point

Tool No.	Description	Aux.	D/1	D/2
0-802FPH1S	2F,5/16TL,1-5/8OAL,PLUG,H1,STF,SPT	TAP	0	80
0-802FBH1S	2F,5/16TL,1-5/8OAL,BOTT,H1,STF,SPT	TAP	0	80
12-282FPH3S	2F,15/16TL,2-3/8OAL,PLUG,H3,STF,SPT	TAP	12	28
12-282FBH3S	2F,15/16TL,2-3/8OAL,BOTT,H3,STF,SPT	TAP	12	28

EXAMPLES: TAPS – Machine Screw 0.002 over H3 limits = H7 Spiral Point

Tool No.	Description	Aux.	D/1	D/2
6-322FPH7S	2F,11/16TL,2OAL,PLUG,H7,STF,SPT,SCRW	TAP	6	32
10-322FPH7S	2F,7/8TL,2-3/8OAL,PLUG,H7,STF,SPT,SC	TAP	10	32
6-323FBH7S	2F,11/16TL,2OAL,BOTT,H7,STF,SPT,SCRW	TAP	6	32
8-322FBH7S	2F,3/4TL,2-1/8OAL,BOTT,H7,STF,SPT,SC	TAP	8	32

Stub Flute Class Place S Before Tool No.
If Spiral Point Modifier Place S At End Of Tool No.
Spiral Point Tap
EXAMPLES : Standard Hand Tap Stub Flute Spiral Point

Tool No.	Description	Aux.	D/1	D/2
S1/4-202FPH3S	2F,1TL,2-1/2OAL,PLUG,H3,STUB,SPT,TAP	TAP	0.25	20
S5/16-182FPH3S	2F,1-1/8TL,2-23/32OL,PLG,H3,STUB,SPT	TAP	0.3125	24
S1/2-133FPH3S	3F,1.66TL,3-3/8OAL,PLUG,H3,STUB,SPT	TAP	0.5	13

EXAMPLES: Standard Hand Tap Screw Machine Taps Stub Flute Spiral Point

Tool No.	Description	Aux.	D/1	D/2
S4-402FPH2S	2F,9/16TL,1-7/8OAL,PLUG,H2,STUB,SPT	TAP	4	40
S12-242FPH3S	2F,15/16TL,2-3/8OAL,PLUG,H3,STUB,SPT	TAP	12	24

TAPS — METRIC

TOOL NO.: TAPSIZE-PITCH/NO.FLUTE/P-B/DTOL./MODIFIERS

Modifiers: Straight Flutes Standard
S-Spiral Point Plug Tap Only

TAP CLASS OF TOOLS: TOOL DESCRIPTION FORMAT
NO. FLUTES, P-B, DTOL, MODIFIER, METRICTAP

Aux. Code Format and Dimension Definition
Aux. I.D.: TAPx D/1: Size D/2: Pitch

EXAMPLES: TAPS – Metric Standard Straight Flutes

Tool No.	Description	Aux.	D/1	D/2
1.6-.352FPD3	2F,PLUG,D3,STRF,METRIC	TAP	1.6	0.35
1.6-.352FBD3	2F,BOTT,D3,STRF,METRIC	TAP	1.6	0.35
1.6-.352FBD3S	2F,BOTT,D3,STRF,SPT,METRIC	TAP	1.6	0.35
20-2.54FPD7	4F,PLUG,D7,STRF,METRIC	TAP	20	2.5
20-2.54FBD7	4F,BOTT,D7,STRF,METRIC	TAP	20	2.5
20-2.54FBD7S	4F,BOTT,D7,STRF,SPT,METRIC	TAP	20	2.5

TAPS — PIPE

TOOL NO.: TAPSIZE-TPI/NPS-NPSF/MODIFIER

NPT-NPTF for taper pipe taps

TAP CLASS OF TOOLS: TOOL DESCRIPTION FORMAT
NO. FLUTES, THRDLNGTH, OAL, MODIFIER,PIPE TAP

Aux. Code Format and Dimension Definition
Aux. I.D.: TAPP D/1: Size D/2: Threads per inch

EXAMPLES: PIPE TAPS – Straight Fractional Plug Style Only

Tool Number	Description	Aux.	D/1	D/2
1/8-27NPS	4F,3/4TL,2-1/8OAL,PIPETAP	TAPP	0.125	27
1/8-27NPSLS	4F,3/4TL,2-1/8OAL,LNG,PIPETAP	TAPP	0.125	27
1-11.5NPS	5F,1-3/4TL,3-3/4OAL,PIPETAP	TAPP	1.0	11.5
1/8-27NPSF	4F,3/4TL,2-1/8OAL,PIPETAP	TAPP	0.125	27
1/8-27NPSFLS	4F,3/4TL,2-1/8OAL,,LNG,PIPETAP	TAPP	0.125	27
3/8-18NPSF	4F,1-1/16TL,2-9/16OAL,PIPETAP	TAPP	0.375	18

EXAMPLES: PIPE TAPS – Taper Pipe Taps;

I-Interrupted; H-Hihook; N-Neutral

Tool No.	Description	Aux.	D/1	D/2
1/8-27NPT	4F,11/16TL,2-1/8OAL,PIPETAP	TAPP	0.125	27
1/8-27NPTF	4F,11/16TL,2-1/8OAL,PIPETAP	TAPP	0.125	27
1/8-27NPTI	4F,11/16TL,2-1/8OAL,INRUP,PIPETAP	TAPP	0.125	27
1/8-27NPTFI	4F,11/16TL,2-1/8OAL,INRUP,PIPETAP	TAPP	0.125	27
1/8-27NPTH	4F,11/16TL,2-1/8OAL,HIHK,PIPETAP	TAPP	0.125	27
1/8-27NPTFH	4F,11/16TL,2-1/8OAL,HIHK,PIPETAP	TAPP	0.125	27
1/8-27NPTN	4F,11/16TL,2-1/8OAL,NEUT,PIPETAP	TAPP	0.125	27
1/8-27NPTFN	4F,11/16TL,2-1/8OAL,NEUT,PIPETAP	TAPP	0.125	27
1/8-27NPTLS	4F,11/16TL,2-1/8OAL,LNG,PIPETAP	TAPP	0.125	27
1/8-27NPTFLS	4F,11/16TL,2-1/8OAL,LNG,PIPETAP	TAPP	0.125	27
1/8-27NPTLSI	4F,11/16TL,2-1/8OAL,LNG,IRUP,PIPETAP	TAPP	0.125	27
1/8-27NPTFLSI	4F,11/16TL,2-1/8OAL,LNG,IRUP,PIPETAP	TAPP	0.125	27
1/8-27NPTLSH	4F,11/16TL,2-1/8OAL,LNG,HIHK,PIPETAP	TAPP	0.125	27
1/8-27NPTFLSH	4F,11/16TL,2-1/8OAL,LNG,HIHK,PIPETAP	TAPP	0.125	27
1/8-27NPTLSN	4F,11/16TL,2-1/8OAL,LNG,NEUT,PIPETAP	TAPP	0.125	27
1/8-27NPTFLSN	4F,11/16TL,2-1/8OAL,LNG,NEUT,PIPETAP	TAPP	0.125	27
1.5-11.5NPT	7F,1-3/4TL,4-1/4OAL,PIPETAP	TAPP	1.5	11.5
1.5-11.5NPTF	7F,1-3/4TL,4-1/4OAL,PIPETAP	TAPP	1.5	11.5
1.5-11.5NPTI	7F,1-3/4TL,4-1/4OAL,INRUP,PIPETAP	TAPP	1.5	11.5
1.5-11.5NPTFI	7F,1-3/4TL,4-1/4OAL,INRUP,PIPETAP	TAPP	1.5	11.5
1.5-11.5NPTH	7F,1-3/4TL,4-1/4OAL,HIHK,PIPETAP	TAPP	1.5	11.5
1.5-11.5NPTFH	7F,1-3/4TL,4-1/4OAL,HIHK,PIPETAP	TAPP	1.5	11.5
1-11.5NPTN	5F,1-3/4TL,3-3/4OAL,NEUT,PIPETAP	TAPP	1.0	11.5
1-11.5NPTFN	5F,1-3/4TL,3-3/4OAL,NEUT,PIPETAP	TAPP	1.0	11.5

TAPS — ACME

TOOL NO.: A/TAPSIZE-TPI/MODIFIERS

Modifiers: Straight Flutes Standard
T-Tandem

ACME TAPS CLASS OF TOOLS: TOOL DESCRIPTION FORMAT
THRDLNGTH, OAL, MODIFIER, ACME TAP

Aux. Code Format and Dimension Definition
Aux. I.D.: TAPA D/1: Size D/2: Threads per Inch

EXAMPLES: ACME TAPS

Tool Number	Description	Aux.	D/1	D/2
A1/2-10	2-9/16TL,5OAL,ACME TAP	TAPA	0.5	10
A1-1/2-4	5-7/8TL,12-5/8OAL,ACME TAP	TAPA	1.5	4

TOOL NUMBER CONSTRUCTION GUIDE — DIES: ROUND ADJUSTABLE & RETHREADING, AND ROUND ADJUSTABLE & HEX RETHREADING

TOOL NO.: TAPSIZE-TPI/MODIFIER/O.D.

SIZE	STYLE	OUTSIDE DIA.	MODIFIER
0-80	**A**	**13/16**	**L**
Tap Size	A	O. D.	L
	Adjustable	13/16	Left Hand
		Threads Per In.	
		0-80	R
		1/4-20	ReThreading
			Repair

DIE CLASS OF TOOLS: TOOL DESCRIPTION FORMAT
DIEx: O.D./ACROSS FLAT,THICKNESS,ADJUSTABLE/RETHRDG DIE

Aux. Code Format and Dimension Definition
Aux. I.D.: DIEx D/1: Size D/2: Threads per Inch
x = A – Adjustable; R - Repair or Rethreading

EXAMPLES

Tool No.	Description	Aux.	D/1	D/2
0-80A13/16	13/16O.D.,1/4THK,ADJUSTABLE DIE	DIEA	0	80
6-32A1	1O.D.,3/8THK,ADJUSTABLE DIE	DIEA	6	32
1/4-20A1.5	1.5O.D.,1/2THK,ADJUSTABLE DIE	DIEA	0.25	20
1/4-20A2	2O.D.,5/8THK,ADJUSTABLE DIE	DIEA	0.25	20
5/8-11A2.5	2.5O.D.,3/4THK,ADJUSTABLE DIE	DIEA	0.625	11
1.5-12A3	3O.D.,1THK,ADJUSTABLE DIE	DIEA	1.5	12
1/8-27A1	1O.D.,ADJUSTABLE PIPE DIE	DIEA	0.125	27
1/8-27A1.5	1.5O.D.,ADJUSTABLE PIPE DIE	DIEA	0.125	27
1/8-27A2	2O.D.,ADJUSTABLE PIPE DIE	DIEA	0.125	27
1/4-20R	19/16FLAT,1/4THK,REPAIR DIE	DIER	0.250	20
1.5-12R	2-9/16FLAT,1THK,REPAIR DIE	DIER	1.5	12
1/4-20A1L	1O.D.,LEFTHAND ADJUSTABLE DIE	DIEA	0.25	20

IV. Boring Tools

Boring always refers to relatively precise machining, even when it is a rough boring operation. Boring can be easy as well as extremely difficult, when the depth and diameter ratios increase, and particularly when diameters decrease.

Boring can be done with solid carbide or HHS tools, with tools that use indexable inserts, and with combinations that are used with proprietary tooling systems. ITCS has also included counterbores, countersinks, and other hole finishing tools in this category. In future editions, other boring tools, such as boring bars, cartridges, boring heads, etc. will be added. For the present, we have found that most customers use the ANSI standards or the manufacturer's number for most boring tools.

Constructing Tool Numbers and Aux. I.D.'s for BORES

Considerations used for General Tool Number Construction

CLASS	TYPE	MODIFIER	MODIFIER
TYPE	FRACTIONAL		
OF	SCREW SIZE NO.		HSS
BORES	STOCK NO.		CARBIDE
COUNTER		W & W/0 PILOTS	
BORES	HEX HEAD		
	FILLISTER	TAPER LENGTH	
	FLAT HEAD		
SPOTFACERS		STRAIGHT SHANK	
PILOTS		INTERCHANGEABLE	
		PILOTS	
BORING			
BITS	STOCK NO.		COBALT
ROTO			
BORES			
COUNTER	FRACTIONAL	60,82,90,100,110,120	HSS
SINKS		INCLUDED ANGLES	CARBIDE

COUNTERBORES & SPOTFACERS, BORING BITS, ROTOBORES, COUNTERSINKS

Tool No. Begins	General Description	TLC Aux. I.D.
CB	Counterbores and Spotfacers - Fractional, or Screw No. Size - With & Without Pilots, Interchangeable Pilots - Straight & Taper Shank, Fillister, Flat, Hex Head & Cap Screws, Square Nuts	CB
BB	Boring Bits	BBIT
RB	Roto Bores	BROT
CS	Countersinks	CSNK
P	Spotface Pilot	P

Notes:

1) Counterbores can be supplied with or without pilots. Pilots can also be interchangeable. Counterbores will be listed without pilots if they are capable of accommodating interchangeable pilots. 2) Boring bits are considered to be cobalt steel standard. Bits come in two lengths — short for rigidity, and long with offset taper. Shanks are assumed to fit standard holders. 3)Roto Bores are for cutting concentric burrless holes in thin stock. They come with straight and taper shank and are specified by stock number. 4) Countersinks can be specified in HSS (considered standard) and carbide. Chatterless or chatterfree is considered standard type.

TOOL NUMBER CONSTRUCTION GUIDE — COUNTERBORES & SPOTFACERS

CB and P - COUNTERBORES AND SPOTFACER PILOTS

TOOL NO.: CB/FRACTION, etc./MODIFIER
TOOL NO.: P/SHANKDIA/PILOTDIA

CLASS	SIZE	MODIFIER
CB	**1/4**	**TS 5/32**
CB	FRACTIONAL	TS
Counterbore	1/4	Taper Shank Straight Shank Standard
	Screw Size 2 or #2	Pilot Dia. 5/32
....	....	....
P Spotface Pilot	Shank Dia 3/32	Pilot Dia. -5/32

CB & P CLASS OF TOOLS: TOOL DESCRIPTION FORMAT

DIA./SCREW NO.,SHANKDIA/TAPER,PILOTHOLE/DIA.

Aux. Code Format and Dimension Definition
Aux. I.D.: [CB,P] D/1: Cut/Pilot Diameter D/2: Shank Diameter

EXAMPLES: COUNTER BORE & PILOT TOOLS

Tool No.	Description	Aux	D/1	D/2
CB1/4	1/4D,3/32HOLE,15/64SHNK,C'BORES'FACE	CB	0.25	0.23437
CB1/4TS	1/4D,3/32HOLE,#1TPRSHNK,C'BORES'FACE	CB	0.25	1.0000
CB1-15/16	1-15/16D,7/16HOLE,1.5SHNK,C'BRS'FACE	CB	1.9375	1.5000
CB1-15/16TS	1-15/16D,7/16HOLE,#4TPR,C'BORES'FACE	CB	1.9375	4.0000
CB3	3D,1/2HOLE,1-3/4SHNK,C'BORE S'FACE	CB	3.0000	1.7500
CB3TS	3D,1/2HOLE,#5TPRSHNK,C'BORE S'FACE	CB	3.0000	5.0000
P3/32-1/8	3/32SHNK,1/8PILOT,S'FACE PILOT	P	0.125	0.09374
P7/16-1-13/16	7/16SHNK,1-13/16PILOT,S'FACE PILOT	P	1.8125	0.4375
P1/2-2	1/2SHNK,2PILOT,S'FACE PILOT	P	2.0000	0.50000
CB7/16-3/16	7/16D,3/8SHNK,3/16PILOT,HEXHDC'BRE	CB	0.4375	0.375
CB1-23/32-7/8	1-23/32D,1SHK,7/8PILOT,HEXHD C'BORE	CB	1.7187	1.0000
CB1-15/16-1	1-15/16D,1SHK,1PILOT,HEXHD C'BORE	CB	1.9375	1.0000
CBNO2	FILLISTER C'BORE 3/16SHNK,.085PILOT	CB	0.155	0.1875
CBNO12	FILLISTER C'BORE 3/8SHNK..215PILOT	CB	0.372	0.3750
CB#2	FLATHD SCRWCBORE 3/16SHNK,.085PILOT	CB	0.172	0.1875
CB#12	FLATHD SCRWCBORE 3/8SHNK..215PILOT	CB	0.438	0.3750
CB#3/8	FLATHD SCRWCBORE 1/2SHNK..374PILOT	CB	0.762	0.3750
CB1.531-1.031	1.531D,1SHNK,1.031PILOT,CAPSCRC'BORE	CB	1.531	1.0000
CB1.344-1.031	1.344D,1SHNK,1.031PILOT,CAPSCRC'BORE	CB	1.344	1.0000
CB1.531-1	1.531D,1SHNK,1.000PILOT,CAPSCRC'BORE	CB	1.531	1.0000
CB1.5-1	1.500D,1SHNK,1.000PILOT,CAPSCRC'BORE	CB	1.500	1.0000
CB1.344-1	1.344D,1SHNK,1.000PILOT,CAPSCRC'BORE	CB	1.344	1.0000
CB1.312-1	1.312D,1SHNK,1.000PILOT,CAPSCRC'BORE	CB	1.312	1.0000
CB3/8-1/4	3/8D,5/16SHNK,1/4PILOT,CAPSCRWC'BORE	CB	0.375	0.3125
CB13/32-1/4	13/32D,5/16SHK,1/4PILOT,CAPSCRC'BORE	CB	0.406	0.3125
CB13/32-9/32	13/32D,5/16SHK,9/32PILOT,CAPSCRC'BOR	CB	0.406	0.3125

TOOL NUMBER CONSTRUCTION GUIDE — BORING BITS & ROTOBORES

BB - BORING BITS & RBS & RBM - ROTOBORES

TOOL NO.: Class Code/STOCKNO./MODIFIER

CLASS	SIZE	MODIFIER
BB	**AH3**	**S**
BB Boring Bits	STOCK NO. AH3	S Short L Long
RBS Straight Shank Rotobore RBM Taper Shank Rotobore	Stock No. 2-1A	

BB, RBS, AND RBM CLASS OF TOOLS: TOOL DESCRIPTION FORMAT

BB: MINBOREDIA,MAXBOREDEPTH, SHNKDIA, BORINGBIT
RBS & RBM: HEADDIA, SHANKDIA, ROTOBORE

Aux. Code Format and Dimension Definition

Aux. I.D.: BBIT D/1: Minimum Bore Dia. D/2: Maximum Bore Depth
BROT D/1: Head Diameter D/2: Shank Diameter

EXAMPLES: BORING BIT & ROTOBORE TOOLS

Tool No.	Description	Aux.	D/1	D/2
BBAH3S	3/16DMIN,9/16DEEP,3/8SHNK,BORINGBIT	BBIT	0.1875	0.5625
BBAH3L	3/16DMIN,15/16DEEP,3/8SHNK,BORINGBIT	BBIT	0.1875	0.9375
BBAH6S	3/8DMIN,1-1/8DEEP,3/8SHNK,BORINGBIT	BBIT	0.375	1.125
BBAH6L	3/8DMIN,1-7/8DEEP,3/8SHNK,BORINGBIT	BBIT	0.375	1.875
BBBH4S	1/4DMIN,3/4DEEP,1/2SHNK,BORINGBIT	BBIT	0.250	0.750
BBBH4L	1/4DMIN,1-1/4DEEP, 1/2SHNK,BORINGBIT	BBIT	0.250	1.250
BBBH8S	1/2DMIN,1-1/2DEEP,1/2SHNK,BORINGBIT	BBIT	0.500	1.500
BBBH8L	1/2DMIN,2-1/2DEEP,1/2SHNK,BORINGBIT	BBIT	0.500	2.500
BBDH6S	3/8DMIN,1-1/8DEEP,5/8SHNK,BORINGBIT	BBIT	0.375	1.125
BBDH6L	3/8DMIN,1-7/8DEEP,5/8SHNK,BORINGBIT	BBIT	0.375	1.875
BBDH10S	5/8DMIN,1-7/8DEEP,5/8SHNK,BORINGBIT	BBIT	0.625	1.875
BBDH10L	5/8DMIN,3-1/8DEEP,5/8SHNK,BORINGBIT	BBIT	0.625	3.125
BBCH6S	3/8DMIN,1-1/8DEEP,3/4SHNK,BORINGBIT	BBIT	0.375	1.125
BBCH6L	3/8DMIN,1-7/8DEEP,3/4SHNK,BORINGBIT	BBIT	0.375	1.875
BBCH12S	3/4DMIN,2-1/4DEEP,3/4SHNK,BORINGBIT	BBIT	0.750	2.250
BBCH12L	3/4DMIN,3-3/4DEEP,3/4SHNK,BORINGBIT	BBIT	0.750	3.750
RBS2-1A	3/16D,1/4SHNK,ROTOBORE	BROT	0.1875	0.25
RBS2-10A	3/4D,1/2SHNK,ROTOBORE	BROT	0.750	0.5
RBS4-11A	13/16D,1/2SHNK,ROTOBORE	BROT	0.8125	0.5
RBS4-29A	1-15/16D,1/2SHNK,ROTOBORE	BROT	1.9375	0.5
RBS4-30A	2D,1/2SHNK, ROTOBORE	BROT	2.000	0.5
RBM4-14A	1D,#2MTSHNK,ROTOBORE	BROT	1.000	2
RBM4-15A	1-1/6D,#2MTSHNK,ROTOBORE	BROT	1.0625	2
RBM4-29A	1-15/16D,#3MTSHNK,ROTOBORE	BROT	1.9375	3
RBM4-30A	2D,#3MTSHNK,ROTOBORE	BROT	2.000	3

TOOL NUMBER CONSTRUCTION GUIDE — COUNTERSINKS

CS - COUNTERSINKS

TOOL NO.: CS/BODYDIA./#FLUTES/INCL.ANGLE/MODIFIER

CLASS	SIZE	# OF FLUTES	INCLD ANGLE	MODIFIER
CS	**1/4**	**6F**	**60**	**CB**
CS	FRACTIONAL	4F	60	CB
Counter	1/4	6F	82	Carbide
Sink	Body Dia.	No. of flutes	90	
		Omit if 1F	100	
			110	
	Optional: Use		120	
	STOCK NO.		Included Angle	
	example:			
	CSJC-1A			

CS CLASS OF TOOLS: TOOL DESCRIPTION FORMAT
CSNK: BODYDIA,SHANKDIA,#FLUTES,INCL.ANGLE, MOD.,C'SINK

Aux. Code Format and Dimension Definition
Aux. I.D.: CSNK D/1: Body Diameter D/2: Included Angle

EXAMPLES: COUNTERSINK TOOLS

Tool No.	Description	Aux.	D/1	D/2
CS3/166F60	3/16D&SHNK,6F,60DEG, C'SINK	CSNK	0.1875	60
CS3/166F60CB	3/16D&SHNK,6F,60DEG,CARBIDE C'SINK	CSNK	0.1875	60
CS1-1/26F120CB	1-1/2D,3/4SHNK,6F,120DEG,CARB,C'SINK	CSNK	1.500	120
CS1-3/46F120	1-3/4D,1SHNK,6F,120DEG, C'SINK	CSNK	1.750	120
CS26F120	2D,1SHNK,6F,120DEG, C'SINK	CSNK	2.000	120
CS1/44F60	1/4D,3/16SHNK,4F,60DEG, C'SINK	CSNK	0.250	60
CS1-3/44F120	1-3/4D,1SHNK,4F,120DEG, C'SINK	CSNK	1.750	120
CS24F120	2D,1SHNK,4F,120DEG C'SINK	CSNK	2.000	120
CS1/44F120CB	1/4D,3/16SHNK,4F,120DEG,CARB, C'SINK	CSNK	0.250	120
CS14F120CB	1D,1/2SHNK,4F,120DEG,CARB, C'SINK	CSNK	1.000	120
CS1/460	1/4D,3/16SHNK,1F,60DEG, C'SINK	CSNK	0.250	60
CS260	2D,1SHNK,1F,60DEG, C'SINK	CSNK	2.000	60
CS1/482	1/4D,3/16SHNK,1F,82DEG, C'SINK	CSNK	0.250	82
CS282	2D,1SHNK,1F,82DEG, C'SINK	CSNK	2.000	82
CS1/490	1/4D,3/16SHNK,1F,90DEG, C'SINK	CSNK	0.250	90
CS290	2D,1SHNK,1F,90DEG, C'SINK	CSNK	2.000	90
Optional:				
CSJC-1A	1/4D,3/16SHNK,1F,60DEG, C'SINK	CSNK	0.250	60
CSJC-10A	2D,1SHNK,1F,60DEG, C'SINK	CSNK	2.000	60
CSJC-1B	1/4D,3/16SHNK,1F,82DEG, C'SINK	CSNK	0.250	82
CSJC-10B	2D,1SHNK,1F,82DEG, C'SINK	CSNK	2.000	82
CSJC-1C	1/4D,3/16SHNK,1F,90DEG, C'SINK	CSNK	0.250	90
CSJC-10C	2D,1SHNK,1F,90DEG, C'SINK	CSNK	2.000	90

V. Milling Cutters

Milling generates a machined surface by removing material from the workpiece with a rotating cutter. The two most basic types of milling are often referred to as face and peripheral, but the differences are not all that great. Other types of milling operations are slitting, endmilling, side and slot milling, slab, straddle, hollow, and gear milling. All milling operations are dictated by the workpiece part and the style of cutter selected and they all can be classified as either a face or peripheral milling operation.

Constructing Tool Numbers and Aux. I.D.'s for MILLS

Considerations used for General Tool Number Construction

TYPE of MILL	SIZE	MODIFIER	MODIFIER
		SINGLE END	
	FRACTIONAL	DOUBLE END	
		EXTENDED SHANK	HSS
	DECIMAL		PREMIUM
STD.		CENTER CUTTING	COBALT
ENDMILLS	MILLIMETER		CARBIDE
			TiN COATED
	DIAMETER	STUB FLUTE	
MINIATURE		SHORT FLUTE	STRAIGHT SHANK
END MILLS			SHANK DEVIATION
		MEDIUM LENGTH	LENGTH OF CUT
BALL		LONG LENGTH	
END MILLS		XTRA LONG LENGTH	HEAVY DUTY
ROUGHING		ALUMINUM CUTTING	B&S TAPER
ENDMILLS			MORSE TAPER
SHELL		STRAIGHT FLUTE	LEFT HAND CUT
MILLS		HIGH SPIRAL	LEFT HAND SPIRAL
HOLLOW			RIGHT HAND CUT
MILLS			RIGHT HAND SPIRAL
TSLOT			
CUTTERS		BOLT SIZE	
TAPERED	STOCK NO.		
ENDMILLS		TAPER ANGLE	
DOVE TAIL			
CUTTERS			
PLAIN			
CUTTERS	FRACTIONAL		HALF SIDE
	DECIMAL		STAGGERED TOOTH
SIDE	MILLIMETER		
MILLS			
SINGLE &	DIAMETER	WIDTH	
DOUBLE		ANGLE	HEAVY DUTY
ANGLE MILL			
			HELIX CHANGES
CONCAVE			
CONVEX			RIGHT HAND CUT
CORNER	RADIUS		LEFT HAND CUT
ROUNDING			
CUTTERS			
WOODRUFF			
KEYSEAT	STOCK NO.		
CUTTERS	STANDARD NO.		

MILLS & CUTTERS – END, BALL, ROUGHING, SHELL, TSLOT, DIE SINKING, HOLLOW, DOVETAIL,CORNER ROUNDING, PLAIN, SIDE, HALF, STAGGERED TOOTH, ANGLE, CONCAVE, CONVEX, CORNER ROUNDING, WOODRUFF KEYSEAT

Tool No. Begins	General Description	TLC Aux. I.D.
E	ENDMILLS - Fractional, Millimeter - Stub Length, Stub Flute, Single End, Double End, Short Flute, Long Flute, Xtra Long Flute, Straight Flute, Center Cutting - HHS, Premium. Cobalt, Carbide, TiN, Aluminum Cut, Right & Left Hand Spiral & Cuts, Low & High Spiral, Heavy Duty, Light Duty, Extended Length, Two Lipped, B&S Taper, Morse Taper	E
ME	Miniature End Mills	EM
BE	Ball End Mill	EB
BME	Miniature Ball End Mills	EMB
RBE	Roughing Ball End Mills	EBR
RE	Roughing End Mills	ER
TE	Tapered Die Sinking End Mills	ET
CRE	Corner Rounding End Mills	ECR
HM	Hollow Mills	MH
SM	Shell Mills	MS
RSM	Roughing Shell Mills	MSR
TSC	T Slot Cutters	CTS
RFTSC	Roughing & Finishing T Slot Cutters	CTSR
DTC	Dove Tail Cutters	CDT
PMC	Plain Milling Cutters	CPM
SMC	Side Milling Cutters	CSM
SSMC	Staggered Tooth Side Milling Cutters	CSSM
HSMC	Half Side Milling Cutters	CHSM
SAMC	Single Angle Milling Cutters	CSAM
DAMC	Double Angle Milling Cutters	CDAM
CCMC	Concave Milling Cutters	CCCM
CVMC	Convex Milling Cutters	CCVM
CRMC	Corner Rounding Milling Cutters	CCRM
WKC	Woodruff Keyseat Cutters	CWK
WNC	Woodruff Narrow Width Cutters	CWN
WKCS	Shank Style Woodruff Keyseat Cutters	CWKS

Notes:

1) Standard endmills have radial grind and double backoff. Endmills under 3/16" diameter have single radial grind. Unless otherwise specified, all endmills are right hand cut, right hand spiral, and high speed steel. Generally, endmills 7/8" diameter and over are provided with 2 driving flats. Unless otherwise specified, right hand mills have right hand flutes and left hand mills have left hand flutes. B&S no. 5 taper shanks have plain ends without tang. B&S no. 7 and 9 taper shanks are provided with tang. Morse taper shanks provided with tang end. 2) Roughing endmills are provided with 30 degree helix except for ¼” and 3/8" diameter which are provided with 25 degree helix. 2" diameter shank roughing endmills have Bayonet type shanks. Roughing shell mills have 30 degree helix. Center cutting roughing endmills have 36 degree helix and are provided undersized. Aluminum center cutting roughing endmills have deeper gullets and polished flutes. 3)Tapered endmills have constant spirals, double backoff, high helix, 3 flutes, end cutting, and triple draw. 4) Milling cutters with face widths up to ¾” have straight teeth, sizes ¾” and over have 18 degree left hand helix. Heavy duty milling cutters have 45 degree helix, 10 degree positive rake, and left hand helix. Helical plain milling cutters have 52 degree left hand helix angle and 10 degree positive rake. Side mills have 10 degree positive rake. Half side mills have 15 degree helix. 5) Woodruff keyseat cutters are provided 1/32" oversized diameter, staggered tooth without relief on side teeth, and of correct dimensions for keyseat stock number. Shanked woodruff and woodruff keyseat cutters are ½” diameter and 2" under the head.

TOOL NUMBER CONSTRUCTION GUIDE — ENDMILLS

E; ME; BE; BME; RE; REF; RBE – ENDMILLS

TOOL NO.: Class Code/DIA./FLUTETYPE/#FLUTES/ENDTYPE/MODIFIERS

CLASS	SIZE	TYPE OF FLUTE	#FLUTES	END TYPE	MODIFIER
E	**1/8**	**S**	**2F**	**D**	**CO**
E	Diameter	OMIT	2F	OMIT	OMIT if HSS
		if Standard		if Single	
Std.	FRACTIONAL		3F		
Endmill	1/8		4F		
		S	6F	D	P
ME	DECIMAL	Stub or	8F	Double	Premium
Miniature	.125	Short Fl			
Endmill				E	CO
	MILLIMETER	M		Extended	Cobalt
BE	9 MM	Medium Long			
Ballend				C	CB
Endmill		L		Center	Carbide
		Long		Cutting	
BME					S
Miniature		X		Rx	a) Straight
Ballend		Xtra Long		Radius	Shank
				R1/8	S(dia.)
* REF		ST			b) if shank <>
Rgh. Fin.		Straight			cut dia.
Endmill					(dia)SD
		A			c) Stub
					Flutes
RBE		Aluminum			w/
Roughing					different
Ballend					Shank
					Dia.s
		H			d) Lngth of
		High Spiral			Cut
					if same
					Shnk

continued on next page

DESCRIPTOR Priorities

Type Of Flute	End Type	Modifier	
1. Lengths	1. Double	1. Mat'ls	H Heavy Duty
2. Spirals	2. Center Cutting	2. Straight Shank	T TiN Coated
3. Alum	3. Radius	3. Shank Dia	#BS B&S Taper
	4. Heavy Duty		#MT Morse Taper
	5. Length of Cut		LS R Hand Cut L H Spiral
	6. Taper		LL L Hand Cut L H Spiral
	7. Spiral		

ENDMILLS CLASS OF TOOLS: TOOL DESCRIPTION FORMAT

DIA., SHNKDIA., L.O.C., MODIFIERS, MILLTYPE

Aux. Code Format and Dimension Definition
Aux. I.D.: [E, EM, EB, EMB, EBR, ER]
D/1: Diameter D/2: Number of Flutes

EXAMPLES: ENDMILLS

Tool No.	Description	Aux.	D/1	D/2
E1/8S2F	1/8D,3/8SHNK,3/16LOC,STUBFLT,ENDMILL	E	0.125	2
E5/32S2FD	5/32D,3/8SHK,15/64LC,DBL,STBF,STDSHK	E	0.15625	2
E1/82F	1/8D,3/8SHK,3/8LC,SINGEND,ENDMILL	E	0.125	2
E1/4A2F	1/4D,3/8SHK,5/8LC,ALUMCUT,SINGEND EM	E	0.25	2
E1/82FP	1/8D,3/8SHK,3/8LC,PREM,SINGEND EM	E	0.125	2
E12FCO	1D&SHK,1-5/8LC,COBALT SINGEND EMILL	E	1.0	2
E7/8A2FS	7/8D&SHK,1-7/8LC,ALUMCUT,SINGEND EM	E	0.875	2
E1/4S2FS	1/4D&SHK,3/8LC,SHRTFL,STRSHK,SEMILL	E	0.25	2
E1-1/4S2F3/4SD	or 1-1/4D,3/4SHK,3/4LC,STUB ENDMILL	E	1.25	2
E1.25S2F.75SD	1.25D,.75SHK,3/4LC,STUB ENDMILL	E	1.25	2
E1/82FE	1/8D,3/8SHK,3/8LC,13/16EXTD,ENDMILL	E	0.125	2
E1-1/2X2F2-1/4	or 1-1/2D,1-1/4SHK,2-1/4LC,XLNG,ENDMILL	E	1.5	2
E1.5X2F2.25	1.5D,1.25SHK,2.25LC,XLONG,SEND ENDML	E	1.5	2
E1/4X2F	1/4D,3/8SHK,1LC,XTRALNG,SING ENDMILL	E	0.25	2
E1/4L3F	1/4D,3/8SHK,1-1/4LC,LONG,SING ENDMLL	E	0.25	3
E1/4AX2F	1/4D,3/8SHK,1-3/4LC,ALUMXLNG,SENDMLL	E	0.25	2
E1-1/2AX2F8	1-1/2D,1-1/4SHK,8LC,ALUMXLNG,SENDMLL	E	1.5	2
E1-1/4AL2F4	1-1/4D&SHK,4LC,ALUMLNG,SING ENDMILL	E	1.25	2
E22FH	2D&SHK,2LC,HVYDTY ENDMILL	E	2	2
E22FH4	2D&SHK,4LC,HVYDTY ENDMILL	E	2	2
E1/42F5BS	1/4D,#5B&S,3/8LC,2LIPPD,ENDMILL	E	0.25	2
E1/42F7BS	1/4D,#7B&S,3/8LC,2LIPPD,ENDMILL	E	0.25	2
E1/24F2MT	1/2D,#2MT,15/16LC,TPRSHK,ENDMILL	E	0.5	4
E1-1/8L6FLS	1-1/8D,1SHK,4LC,LNG,LHSPRL,ENDMILL	E	1.125	6
E1/84FDLL	1/8D,3/8SHK,3/8LC,DBL,LHCUTLHSPRL,EM	E	0.125	4
E1/82FDCO	1/8D,3/8SHK,3/8LC,DBLEND,COBALT,EMIL	E	0.125	2
E1/82FD	1/8D,3/8SHK,3/8LC,DBLEND,ENDMILL	E	0.125	2
E1/4ST2FD	1/4D,3/8SHK,1/2LC,STRTFL,DBLENDMILL	E	0.25	2
E1-1/4ST2F	1-1/4D,7/8SHK,1-5/8LC,STRTFL,ENDMILL	E	1.25	2
E1/44FCT	1/4D,3/8SHK,5/8LC,CNTRCT,TiN,ENDMILL	E	0.25	4
E1/4A2FT	1/4D,3/8SHK,5/8LC,ALUMCT,TiN,ENDMILL	E	0.25	4
ME1/32S2F	1/32D,3/16SHK,3/64LC,STBFL,MINIAEMIL	EM	0.03125	2
ME1/322F	1/32D,3/16SHK,3/32LC,MINIATURENDMILL	EM	0.03125	2
ME1/322FD	1/32D,3/16SHK,3/32LC,DBLENDMINIATEML	EM	0.03125	2
ME1/16L2FD	1/16D,3/16SHK,7/32LC,LNG,DBL,MIN,EML	EM	0.0625	2
ME1/32S2FD	1/32D,3/16SHK,3/64LC,STBFL,DBL,MINEM	EM	0.03125	2
ME1/16S3FC	1/16D,1/4SKH,3/32LC,SHRT,CNTRCT,MINI	EM	0.0625	3
ME1/16L3FC	1/16D,1/4SKH,3/32LC,LONG,CNTRCT,MINI	EM	0.0625	3
BE1/4LA2F	1/4D,3/8SHK,1-1/4LC,LNG,ALUM,BALLEND	EB	0.25	2
BE1/4XA2F	1/4D,3/8SHK,1-3/4LC,XLNG,ALUM,BALLEM	EB	0.25	2
BE1/4A2F	1/4D,3/8SHK,5/8LC,ALUMCUT,BALLENDMIL	EB	0.25	2
BE7/82F	7/8D,3/4SHK,2LC,BALLENDMILL	EB	0.875	2
BE7/82F7/8SD	7/8D,7/8SHK,2LC,BALLENDMILL	EB	0.875	2
BE1-1/42FE	1-1/4D&SHK,2LC,5EXTND,BALLENDMILL	EB	1.25	2
BE1/4H2FP	1/4D,3/8SHK,7/16LC,HISPRL,PREM,BEM	EB	0.25	2

continued on next page

Tool No.	Description	Aux.	D/1	D/2
BE1/82FD	1/8D,3/8SHK,3/8LC,DBLEND,BALLENDMILL	EB	0.125	2
BE1-1/4L3F	1-1/4D&SHK,4LC,LONG,BALLENDMILL	EB	1.25	3
BE1/4H4FCP	1/4D,3/8SHK,5/8LC,HISPRL,CNTRCT,PREM	EB	0.25	4
BE3/44F3-3/4	3/4D&SHK,3-3/4LC,BALLENDMILL	EB	0.75	4
BME1/322F	1/32D,3/16SHK,3/64LC,MINIAT,BALLENDM	EBM	0.03125	2
BME1/32L2F	1/32D,3/16SHK,3/32LC,LNG,MINI,BALLEM	EBM	0.03125	2
BME1/32S2FD	1/32D,3/16SHK,3/64LC,SHRT,DBL,MINI,B	EBM	0.03125	2
BME1/322FD	1/32D,3/16SHK,3/32LC,DBL,MINI,BALLEM	EBM	0.03125	2
RE1-1/46FS	1-1/4D&SHK,2LC,RGH ENDMILL	ER	1.25	6
RE1-1/46F3/4SD	1-1/4D,3/4SHK,1-1/2LC,RGH ENDMILL	ER	1.25	6
RE2X8F10	2D&SHK,10LC,XLNG,RGH ENDMILL	ER	2	10
RE1-1/2S6F	1-1/2D,3/4SHK,3/4LC,STUB,RGH ENDMILL	ER	1.5	6
RE1-1/43FCS	1-1/4D&SHK,2LC,CNTRCT,RGH ENDMILL	ER	1.25	3
RE1-1/4A3FCS6	1-1/4D&SHK,6LC,ALUMCT,CNTRCT,RGHENDM	ER	1.25	3
RBE1/24FS	1/2D&SHK,1-1/4LC,RGH BALL ENDMILL	EBR	0.5	4
RE1/24FS	1/2D&SHK,1-1/4LC,RGH&FIN ENDMILL	ER	0.5	4

TOOL NUMBER CONSTRUCTION GUIDE — SHELL & HOLLOW MILLS

HM, SM, RSM MILLS

TOOL NO.: Code Class/DIA./MODIFIER

CLASS	SIZE	MODIFIER
SM	**2-1/2**	**LL**
SM Shell Mill	DIAMETER Fractional 2-1/2	LL Left Hand Cut Left Hand Spiral
RSM Roughing Shell Mill		
HM Hollow Mill		

SM, RSM, HM CLASS OF TOOLS: TOOL DESCRIPTION FORMAT
DIA., SHNKDIA., L.O.C., NO.TTH/FLUTES, MODIFIERS

Aux. Code Format and Dimension Definition
Aux. I.D.: [MS,MSR,MH] D/1: Diameter D/2:

EXAMPLES: SHELL MILLS

Tool No.	Description	Aux.	D/1	D/2
HM3/32	3/32D,5/8SHK,3TEETH HOLLOW MILL	MH	0.09375	
HM1	1D,1-3/4SHK,4TEETH,HOLLOW MILL	MH	1.0	
SM1-1/4	1-1/4D,1/2HOLE,1LOC,8TTH,SHELLMILL	MS	1.25	
SM6	6D,2HOLE,2-1/4LOC,16TTH,SHELLMILL	MS	6	
SM2-1/2LL	2-1/2D,1HOLE,1-5/8LC,10TTH,LHCSPRL	MS	2.5	
SM6LL	6D,2HOLE,2-1/4LC,16TTH,LHCSPRL,SM	MS	6	
RSM2	2D,3/4HOLE,1-3/8LC,8FL,RGHSHELLMILL	MSR	2	
RSM6	6D,2HOLE,2-1/4LC,14FL,RGHSHELLMILL	MSR	6	

TOOL NUMBER CONSTRUCTION GUIDE — T SLOT CUTTERS, DIESINKING MILLS, & CORNER ROUNDING MILLS

TSC RFTSC - T SLOT CUTTERS
TOOL NO.: Code Class/BOLT SIZE

TE – TAPERED ENDMILL DIE SINKING
TOOL NO.: TE/STOCK NO.

CRE – CORNER ROUNDING ENDMILLS
TOOL NO.: CRE/RADIUS/ SHANK

CLASS	SIZE	MODIFIER
CRE CRE Corner Round Endmill	**1/16** RADIUS Fractional 1/16	**-3/8** SHANK Fractional -3/8
TSC T Slot Cutter	BOLT SIZE 1/2	
RFTSC Rgh. & Fin. T Slot Cutter		
TE Tapered or Diesinking Endmill	STOCK NO.	

TSC, RFTSC, TE, CRE CLASS OF TOOLS: TOOL DESCRIPTION FORMAT

TSC & RFTSC: TBOLTSIZE, SHANKDIA, MILLTYPE

Aux. Code Format and Dimension Definition
Aux. I.D.: [CTS,CTSR] D/1: Bolt Size D/2:
TE: TIP, L.O.C., SHK, MILLTYPE
Aux. Code Format and Dimension Definition
Aux. I.D.: ET D/1: Length of Cut D/2: Angle/Side
CRE: RADIUS, SHANK, CORNER ROUNDING ENDMILL
Aux. Code Format and Dimension Definition
Aux. I.D.: ECR D/1: Radius D/2:

EXAMPLES: ENDMILLS

Tool No.	Description	Aux.	D/1	D/2
TSC1/4	1/4BOLT,1/2SHK,8TTH,T SLOT CUTTER	CTS	0.25	
TSC1-1/2	1-1/2BOLT,1-1/4SHK,12TTH,TSLOTCUTTER	CTS	1.5	
RFTSC1/2	1/2BOLT,3/4SHK,RGHFIN TSLOT CUTTER	CTSR	0.5	
RFTSC1-1/2	1-1/2BOLT,1-1/4SHK,RGHFINTSLOTCUTTER	CTSR	1.5	
TE1/2B1	1/8TIP,3/8LC,3/8SHK,.5DEG,TAPEREDEND	ET	0.375	0.5
TE1/2D3	1/4TIP,3/4LC,3/8SHK,.5DEG,TAPEREDEND	ET	0.75	0.5
TE1/2D9	1/4TIP,2-1/4LC,3/8SHK,.5DEG,TPRDEND	ET	2.25	0.5
TE10A6	3/32TIP,1-1/2LC,1/2SHK,10DEG,TPRDEND	ET	0.09375	10
CRE1/16-3/8	1/16R,3/8SHK,CORNER ROUNDING ENDMILL	ECR	0.0625	
CRE1/2-3/4	1/2R,3/4SHK,CORNER ROUNDING ENDMILL	ECR	0.5	
CRE1/2-1	1/2R,1SHK,CORNER ROUNDING ENDMILL	ECR	0.5	

TOOL NUMBER CONSTRUCTION GUIDE — DOVETAIL, SINGLE & DOUBLE ANGLE MILLING CUTTERS

DTC, SAMC, DAMC DOVETAIL, & ANGLE MILLING CUTTERS

TOOL NO.: Class Code/DIA./ANGLE/MODIFIER

CLASS	SIZE	ANGLE	MODIFIER
SAMC	**2-3/4**	**60**	**LH**
SAMC Single Angle Milling Cutter	DIAMETER Fractional 2-3/4	ANGLE Degree 60	LH Left Hand Cut
DAMC Double Angle Milling Cutter			
DTC Dovetail Cutter			

SAMC,DAMC, DTC CLASS OF TOOLS: TOOL DESCRIPTION FORMAT
DIA.,CUTWITH/THK, ANGLE, MODIFIER, MILLTYPE

Aux. Code Format and Dimension Definition
Aux. I.D.: [CDT, CSAM, CDAM] D/1: Diameter D/2: Angle

EXAMPLES

Tool No.	Description	Aux.	D/1	D/2
DTC1/260	1/2D,7/32CW,60DEG,DOVETAILCUTTER	CDT	0.5	60
DTC1/245	1/2D,1/8CW,45DEG,DOVETAILCUTTER	CDT	0.5	45
SAMC2-3/445	2-3/4D,1/2THK,45DEG,SINGANGMILLCUT	CSAM	2.75	45
SAMC2-3/460	2-3/4D,1/2THK,60DEG,SINGANGMILLCUT	CSAM	2.75	60
SAMC2-3/445LH	2-3/4D,1/2THK,45DEG,LHSINANGMILLCT	CSAM	2.75	45
SAMC2-3/460LH	2-3/4D,1/2THK,60DEG,LHSINANGMILLCT	CSAM	2.75	60
DAMC2-3/445	2-3/4D,1/2W,45DEG,DBLANG MILLCUT	CDAM	2.75	45
DAMC2-3/460	2-3/4D,1/2W,60DEG,DBLANG MILLCUT	CDAM	2.75	60
DAMC2-3/490	2-3/4D,1/2W,90DEG,DBLANG MILLCUT	CDAM	2.75	90

TOOL NUMBER CONSTRUCTION GUIDE — PLAIN, SIDE, STAGGERED TOOTH, HALF SIDE, WOODRUFF KEYSEAT & NARROW MILLING CUTTERS

PMC, SMC, SSMC, HSMC, WKC, WCN, WKCS MILLING CUTTERS

TOOL NO.: Class Code/WIDTH or STOCK NO./DIA/MODIFIER

CLASS	SIZE	MODIFIER
PMC	**2W 2-1/2**	**HD**
PMC Plain Mill Cutter	WIDTH andDIAMETER Fractional 2W 2-1/2	HD Heavy Duty
SMC Side Mill Cutter		LH Left Hand Cut HX Helix
SSMC Staggered Tooth Milling Cutter		
HSMC Half Side Milling Cutter		
WKC Woodruf Keyseat Cutter		
WCN Woodruff Narrow Cutter		
WKCS Woodruff Keyseat Shanked Cutter		

PMC, SMC, SSMC, HSMC, WKC, WCN, WKCS CLASS OF TOOLS: TOOL DESCRIPTION FORMAT
WIDTH,DIA.,MODIFIERS,MILLTYPE

Aux. Code Format and Dimension Definition

Aux. I.D.: [CPM, CSM, CSSM, CHSM, CWK, CWN, CWKS]

D/1: Width D/2: Diameter

EXAMPLES

Tool No.	Description	Aux.	D/1	D/2
PMC3/16W2-1/2	3/16W,2-1/2D,PLAIN MILL CUTTER	CPM	0.1875	2.5
PMC1.5W2.5	1.5W,2.5D,PLAIN MILL CUTTER	CPM	1.5	2.5
PMC2W2-1/2HD	2E,2-1/2D,HVYDTY,PLAIN MILL CUTTER	CPM	2.0	2.5
PMC2W3HXX	2W,3D,52DEGHELIX,PLAIN MILL CUTTER	CPM	2.0	3.0
SMC3/16W2	3/16W,2D,SIDE MILL CUTTER	CSM	0.1875	2.0
SMC5/16W2-1/2	5/16W,2-1/2D,SIDE MILL CUTTER	CSM	0.3125	2.5
SSMC1/4W2-1/2	1/4W,2-1/2D,STAGGERED TOOTH SIDE MIL	CSSM	0.25	2.5
SSMC5/16W2-1/2	5/16W,2-1/2D,STGGRD TTH,SIDE MILL	CSSM	0.3125	2.5
HSMC3/4W4	3/4W,4D,HALF SIDE MILL	CHSM	0.75	4.0
HSMC3/4W4LH	3/4W,4D,LFTHND CUT SIDE MILL	CHSM	0.75	4.0
WKC26	3/16W,2-1/8D,WOODRUFF KEYSEAT CUTTER	CWK	0.1875	2.125
WCN.02W1/2	.02W,1/2D,WOODRUF NARROW 1/2SHK CUTR	CWN	0.02	0.5
WCN11/64W1-1/8	11/64W,1-1/8D,WOODRUF NRRW 1/2SHK CT	CWN	0.17188	1.125
WKCS202	1/16W,1/4D,WOODRUFF SHNKD KEYSEAT CT	CWKS	0.0625	0.25

TOOL NUMBER CONSTRUCTION GUIDE — CONCAVE, CONVEX, & CORNER ROUNDING MILLING CUTTERS

CCMC; CVMC; CRMC – MILLING CUTTERS

TOOL NO.: Class Code/CIRCLE DIA or RADIUS

CLASS	SIZE
CCMC	**1/8**
CCMC Concave Milling Cutter	DIAMETER Fractional 1/8
CVMC Convex Milling Cutter	
CRCM Corner Rounding Milling Cutter	

CCMC,CVMC,CRMC CLASS OF TOOLS: TOOL DESCRIPTION FORMAT

CIRCLE DIA. or RADIUS,MILLTYPE

Aux. Code Format and Dimension Definition
Aux. I.D.: [CCCM, CCVM] D/1: Circle Diameter D/2:
Aux. I.D. CCRM D/1: Radius D/2:

EXAMPLES

Tool No.	Description	Aux.	D/1	D/2
CCMC1/8	1/8CD,2-1/4D,CONCAVE MILLING CUTTER	CCCM	0.125	
CVMC1/8	1/8CD,2-1/4D,CONVEX MILLING CUTTER	CCVM	0.125	
CRCM1/8	1/8R,2-1/5D,CORNER RADIUS MILLING CT	CCRM	0.125	

VI. Broaching Tools

Generally, broach tooling is specialty tooling, and is engineered to your specifications. ITCS only considers standard common keyway type broaches and their components.

Most broaches are made of HSSs such as, M2, M3, T15, and M42. Carbide tipped broaches are sometimes used on cast iron, but seldom on steel parts or forgings because of rigidity and chipping.

In general, if the material is machinable, it is broachable, and it is better to pull a broach than push it. Pulling loads the broach with tension. A good rule of thumb is that tool length can be 75 times the diameter when pulling and 25 times the diameter when pushing.

Constructing Tool Numbers and Aux. I.D.'s for BROACHES

Considerations Used for General Tool Number Construction

TYPE	FRACTIONAL	MODIFIER	MODIFIER
OF	LETTER		
BROACH			
		PUSH-TYPE	ONE PASS FINISH
KEYWAY			
		PLAIN BUSHINGS	
SQUARE		COLLARED BUSHINGS	
HEXAGON			
ROUND		SHIMS	

BROACHES

Tool No. Begins	General Description	TLC Aux. I.D.
KBR	Keyway Broaches	BRK
BRBC	Collared Bushings for Keyway Broaches	BRBC
BRBP	Plain Bushings for Keyway Broaches	BRBP
BRSH	Shims for Keyway Broaches	BRSH
BRSQ	Square Broaches	BRSQ
BRHX	Hexagon Broaches	BRHX
BRRD	Round Broaches	BRRD

TOOL NUMBER CONSTRUCTION GUIDE — COUNTERBORES & SPOTFACERS, BROACHES, BUSHINGS, SHIMS

TOOL NO.: Class Code/FRACTIONAL or LETTER/MODIFIER

CLASS	SIZE	MODIFIER
KBR KBR Keyway Broach	**1/8** FRACTIONAL Keyway Width 1/8	**A** Broach Std. Letter A
BRBC BRBP Bushings Collared Plain	Diameter 1/2	Broach Std. Letter A or Broach Width -1/8
BRSH Broach Shim	Broach Width or Std. Letter A	Thickness .031
....	....	....
BRSQ Square Broach BRHX Hexagon Broach BRRD Round Broach	FRACTIONAL	

BROACH CLASS OF TOOLS: TOOL DESCRIPTION FORMAT
WIDTH, LETTER, DIMS, TYPE OF BROACH

Aux. Code Format and Dimension Definition
Aux. I.D.: KBR D/1: Width D/2:
[BRBC,BRBP] D/1: Diameter D/2: Length
BRSH D/1: Keyway Width D/2: Thickness
BRSQ – Square; BRHX – Hexagon; BRRD – Diameter

EXAMPLES: BROACHES

Tool No.	Description	Aux.	D/1	D/2
KBR1/16A	1/16W,A,1/8x5", KEYWAY BROACH	KBR	0.0625	
BRBC1/4A	1/4D,A,1-1/8LNG,COLLARED BUSHING	BRBC	0.25	1.1250
BRBP1-7/16D	1-7/16D,D,4LONG,PLAIN BUSHING	BRBP	1.4375	4.000
BRSHA.031	1/8W,A,.031THK,KEYWAYBROACH SHIM	BRSH	0.125	0.031
BRSQ1/8	1/8SQUARE BROACH	BRSQ	0.125	

VII. Saws and Saw Blades

Sawing can be considered a subset of milling. When sawing you must choose the proper blade with consideration of its rake angle, number of teeth to be in the cut, clearance angle, tooth style, and blade material. Circular sawing is similar to slot milling, and some suppliers include circular saw blades in the milling section of their catalog.

Band saw blades are usually constructed from basic steel materials. Carbon-steel blades, with a flexible back or a hard back, are generally the least expensive. Alloy blades, for more vibration and shock resistance, cut steel faster than carbon-steel blades and are less costly than HSS blades. HSS blades were developed over thirty years ago, and have been developed to the point where they can saw faster (6 times) and last longer than standard flexible back carbon-steel blades. The fourth type of blade is the welded edge blade where only the tip of each tooth is hardened, to as much as Rc 63, generally by electron-beam welding systems.

Generally, saw blades have teeth and cut chips, but there are a number of blades that can be used in sawing machines that do not have teeth. These special-purpose blades can be abrasive coated, wire, and knife-edge, to name a few.

Basically, all blades share five characteristics: width, gage, toothform, pitch, and set.

Constructing Tool Numbers and Aux. I.D.'s for SLOTTING & SLITTING SAWS

Considerations used for General Tool Number Construction

SSP	-Plain			
	-Slitting/Slotting	FACE DIA.	WIDTH	*
SSC	w/Side Chip Clearance			
SSCS	& Staggered Teeth			
SSS	-Screw Slotting	WIRE NO.	HOLE DIA.	NO. OF TEETH

SLITTING/SLOTTING SAW TYPES

Tool No. Begins	General Description	TLC Aux. I.D.
SSP	Plain Metal Slitting Saw	SS
SSC	Metal Slitting Saw w/ Side Chip Clearance	SS
SSS	Screw Slotting Saw - Wire Gage No.	SS
SSCS	Metal Slitting Saw w/ Side Chip Clearance & Staggered Teeth	SS

Notes: * Hole diameter will be specified to differentiate between similar saws to prevent duplicate numbers. Dashes (-'s) will be used to separate fields for clarity, i.e., face diameter from width and wire gage number from hole diameter. If necessary, omit dashes to prevent tool number from exceeding 14 characters.

1) Slitting/slotting saws are generally used for ordinary slotting and cut-off operations. They are ground concave on both sides for clearance, have teeth with 0 to 10 degree positive rake angles. 2) Other styles have 10 degree undercut's and straight gash teeth and relieved for side cutting. 3) Saws are made with standard keyways and should be driven by the key extending into collars on both sides.

TOOL NUMBER CONSTRUCTION GUIDE — SLITTING SAWS

SSP - Plain Metal Slitting Saws, SSC - Metal Slitting Saws w/ Side Chip Clearance, SSCS - Metal Slitting Saws w/Side Chip Clearance & Staggered Teeth

TOOL NO.: Class Code/FACEDIA./WIDTH/* HOLEDIA.
(*If duplicates exist use hole diameter to identify.)

CLASS	SIZE	MODIFIER
SSP SSP Slitting Saw Plain	**5 D 1/8** Face Dia. D Width 5D1/8	**1.25** Hole Dia. 1.25
SSC Side Chip Clearance		
SSCS Side Chip Clearance Staggered Tooth		

SSP, SSC, SSCS CLASS OF TOOLS: TOOL DESCRIPTION FORMAT
FACEDIA.,WIDTH,HOLEDIA.,NO.OFTEETH

Aux. Code Format and Dimension Definition
Aux. I.D.: SS D/1: Width D/2: Hole Diameter

EXAMPLES: SLITTING SAWS

Tool No.	Description	Aux.	D/1	D/2
SSP5D1/8	5D,1/8W,1HOLE,44TEETH.PLAINHSS	SS	0.125	1
SSP5D1/81.25	5D,1/8W,1.25HOLE,44TEETH,PLAINHSS	SS	0.125	1.25
SSC8D1/8	8D,1/8W,1HOLE,48TTH.SIDECHIPCLR	SS	0.125	1
SSC8D1/81.25	8D,1/8W,1.25HOLE,48TTH,SIDECHIPCLR	SS	0.125	1.25
SSCS6D3/16	6D,3/16W,1HOLE,40TTH.STGGRDCHPCLR	SS	0.1875	1
SSCS6D3/161.25	6D,3/16W,1.25HOLE,40TTH,STGGRDCHPCL	SS	0.1875	1.25

TOOL NUMBER CONSTRUCTION GUIDE — SLOTTING SAWS

SSS – Screw Slotting Saws – Wire Gage No.'s

TOOL NO.: SSS/WIREGAGENO.,-HOLEDIA., NO.OFTEETH

CLASS	SIZE	MODIFIER
SSS	**24 -5/8**	**60**
SSS	Wire Gage No.-Hole Dia.	# of Teeth
Slotting Saw	24-5/8	60

SSS CLASS OF TOOLS: TOOL DESCRIPTION FORMAT
WGNO.,HOLEDIA,SAWDIA,NO.OFTEETH,SCREWSLOT

Aux. Code Format and Dimension Definition
Aux. I.D.: SS__ D/1: Width D/2: Hole Diameter

EXAMPLES: SLOTTING SAWS

Tool No.	Description	Aux.	D/1	D/2
SSS24-5/860	24(.020),5/8H,2-1/4D,60TTH,SCREWSLOT	SS	0.020	0.625
SSC24-5/890	24(.020),5/8H,1-3/4D,90TTH,SCREWSLOT	SS	0.020	0.625

Constructing Tool Numbers and Aux. I.D.'s for BAND SAW BLADES, POWER HACKSAW BLADES, & SEGMENTED BAND FILES

Considerations used for General Tool Number Construction

TYPE OF BLADE	WIDTH OF BLADE	PITCH NO. OF TEETH SING. BEVEL DBL BEVEL GAGE	* DIFF. FOR DUPLICATES TOOTH FORM EDGE TYPE SET TYPE GAGE

BAND SAW BLADE, POWERHACK SAW & BAND FILE TYPES

Tool No. Begins	General Description	TLC Aux. I.D.
BS	General Purpose Cutoff Blades	BS
BSB	General Purpose Band Saw Blades	BS
BSC	Cobalt or Carbide Band Saw Blades	BS
BSCD	Cobalt Blades Decimal Pitch	BS
BSF	Friction Style Blade w/Hardedge Back	BS
BSH	Honeycomb Style Band Saw Blade	BSH
BSK	Knife Edge Style Band Saw Blade	BSK
BSNW	Narrow Woodworking Band Saw Blade	BSNW
BSSE	Scallop Edge Band Saw Blade	BSS
BSST	Spring Tempered Band Saw Blade	BSST
BST	Tungsten Carbide Band Saw Blade	BST
BSV	Vtooth Style Band Saw Blade	BSV
BSW	Wavy Edge Band Saw Blade	BSW
BF	Segmented Band File Blades	BF
HS	Power Hack Saw Blades	HS

Notes:

* To distinguish between potential duplicate tool numbers, you can generally use either the tooth form, set type or the gage diameter to make the tool number unique. -'s will be used to separate fields for clarity; generally, -'s are best not used unless necessary.

Basic Definitions of Blade Terminology. WIDTH measures the distance from the tooth tip to the back of the blade. Use the widest blade possible for straight cuts and widest that will cut the radii required for contour cuts. PITCH is the number of teeth per inch. To cut thin materials use more teeth, thicker materials use less teeth. SET indicates the amount the teeth are offset to provide side clearance for the blade. The basic sets are RAKER for constant cross section, WAVE for varying cross sections, and STRAIGHT for free cutting materials, i.e., wood or aluminum. GAGE is thickness of the blade. The heavier the gage, the greater the beam strength. TOOTH FORM specifies the shape of the tooth and gullet. Three basic forms are referenced in this coding standard and they are registered trademarks of DoAll Company. They are: PRECISION, the most versatile with a 0 degree rake angle and smooth radius gullet; BUTTRESS, which is like PRECISION with a lengthened gullet; and CLAW TOOTH, which has a 10 degree positive rake and a lengthened gullet. DECIMAL PITCH blades (also by DoAll Company), are also referenced. Decimal Pitch blades are quieter and have longer life blades and effective feature tooth designs that vary tooth pitch and gullet between sectors of the blade.

TOOL NUMBER CONSTRUCTION GUIDE — BAND SAW BLADES

BS – GENERAL PURPOSE CUTOFF; **B**SB – GENERAL PURPOSE; **B**SC – COBALT; **B**SCD – DECIMAL PITCH COBALT; **B**SF – FRICTION HARD EDGE FLEXIBLE BACK CARBON ALLOY

TOOL NO.: Class Code/WIDTH/PITCH/*MODIFIER
*[BS, BSB] If a duplicate exists add the tooth form (Precision, Claw, Buttress) or the Set Type (Raker, Wave, Straight) or the Gage Dim.
*[BSCD] If a duplicate exists add the Gage Dim.

CLASS	SIZE	MODIFIER
BSB	**1/2 6**	**PRC**
BS Gen. Purp. Cut Off Blades	Width & Pitch 1/26	Tooth Form PRC - Precision CLW - Claw BUT - Buttress
BSB Gen. Purp Band Saw Blades		Set Type RKR - Raker WAVE STR - Straight
BSC Cobalt or Carbide Band Saw Blades		Gage Dim. .035 or G to indicate diff. or larger blade
BSCD Cobalt Blades Decimal Pitch		
BSF Friction Blade Hard Edge Back		

BS, BSB, BSC, BSCD, BSF CLASS OF TOOLS: TOOL DESCRIPTION FORMAT
BS: WIDTH, PITCH, GAGE, NOMSET, SETTTYPE, TOOTHFORM BSB: WIDTH, GAGE, PITCH, TOOTHFORM, NOMSET, SETTYPE BSC & BSCD: WIDTH, PITCH, GAGE, NOMSET, SETTTYPE, TOOTHFORM, COBALT BSF: WIDTH, GAGE, PITCH, TOOTHFORM, NOMSET, SETTYPE, FRHEBS

Aux. Code Format and Dimension Definition
Aux. I.D.: BS D/1: Width D/2: Pitch

EXAMPLES: BAND SAW BLADES

Tool No.	Description	Aux.	D/1	D/2
BS1/210	1/2W,10P,.025G,.044NOM,RAKER,PRECNAR	BS	0.5	10
BS1/210.035	1/2W,10P,.035G,.058NOM,RAKER,PRECNAR	BS	0.5	10
BS3/410	3/4W,10P,.035G,.058NOM,RAKER,PRECWID	BS	0.75	10
BS3/410WAVE	3/4W,10P,.025G,.063NOM,WAVE,PRECWIDE	BS	0.75	10
BSB1/26	1/2W,.025G,6P,CLW,.043NOM,RKR,BGPSAW	BS	0.5	6
BSB1/26PRC	1/2W,.025G,6P,PRC,.043NOM,RKR,BGPSAW	BS	0.5	6
BSB1/210	1/2W,.025G,10P,PRC,.043NOM,RKR,BGPSA	BS	0.5	10
BSC1-1/26	1-1/2W,6P,.05G,.08NOM,RAKER,PRCWD,CB	BS	1.5	6
BSCD22-3.1	2W,2-3.1P,05G,.08NOM,STRAKE,PRW,COBL	BS	2	203.1
BSCD22-3.1.063	2W,2-3.1P,.063G,.1NOM,STRAKE,PRW,COB	BS	2	203.1
BSF3/410	3/4W,.035G,10P,PRC,.066NOM,RKR,FRHEB	BS	0.75	10

TOOL NUMBER CONSTRUCTION GUIDE — BAND SAW BLADES

BSK – KNIFE EDGE SINGLE OR DOUBLE BEVEL; BSSE – SCALLOP EDGE SINGLE OR DOUBLE BEVEL; BSW – WAVY EDGE SINGLE OR DOUBLE BEVEL

TOOL NO.: Class Code /WIDTH/BEVEL/GAGE/*MODIFIER
*[BSSE] if a duplicate exists add the Edge Type, usually H for hard.

CLASS	SIZE			MODIFIER
BSSE	**3/8**	**D**	**.020**	**H**
BSK	WIDTH	BEVEL	GAGE	EDGE TYPE
Knife Edge	Fractional	S	Decimal	H - Hard Edge
Band Saw Blade	3/8	Single	.020	
BSSE	D			
Scallop Edge	Double			
Band Saw Blade				
	DD			
BSW	Double Bevel			
Wavy Edge	Double Edge			
Band Saw Blade				

BSK, BSSE, BSW Class of Tools: Tool Description Format
WIDTH,BEVEL,GAGE

Aux. Code Format and Dimension Definition
Aux. I.D.: [BSK, BSS, BSW] D/1: Width D/2: Gage

EXAMPLES: BAND SAW BLADES

Tool No.	Description	Aux.	D/1	D/2
BSK3/8S.020	3/8W,SB,.020G,STRAIGHTKNIFEEDGE	BSK	0.375	0.02
BSK3/8D.020	3/8W,DB,.020G,STRAIGHTKNIFEEDGE	BSK	0.375	0.02
BSK3/8DD.020	3/8W,DDB,.020G,STRAIGHTKNIFEEDGE	BSK	0.375	0.02
BSSE3/8S.020	3/8W,SB,.020G,REGEDG,SCALLOPEDGEBS	BSS	0.375	0.02
BSSE3/8S.020H	3/8W,SB,.020G,HRDEDG,SCALLOPEDGEBS	BSS	0.375	0.02
BSSE3/8D.020H	3/8W,DB,.020G,HRDEDG,SCALLOPEDGEBS	BSS	0.375	0.02
BSSE3/8D.020	3/8W,DB,.020G,REGEDG,SCALLOPEDGEBS	BSS	0.375	0.02
BSW1-1/4S.025	1-1/4W,SNGL,.025GAGE,WAVYEDGE BS	BSW	1.25	0.025
BSW1-1/4D.025	1-1/4W,DBL,.025GAGE,WAVYEDGE BS	BSW	1.25	0.025

TOOL NUMBER CONSTRUCTION GUIDE — BAND SAW BLADES

BSH – HONEYCOMB ALLOY TEMPERED STEEL BACK; BSST – SPRING TEMPERED; BSNW – NARROW WOODWORKING

TOOL NO.: Class Code/WIDTH/GAGE/*TEETHPERIN.
* [BSNW, BSST] add teeth per inch.

CLASS	SIZE		MODIFIER
BSNW	**3/16**	**.028**	**4**
BSH Honeycomb Band Saw Blade	WIDTH Fractional 3/16	GAGE Decimal .028	TEETH PER IN. 4
BSNW Narrow Woodworking Band Saw Blade			
BSST Spring Tempered Band Saw Blade			

BSH, BSNW, BSST Class of Tools: Tool Description Format
BSH: WIDTH, GAGE, TEETHPERIN., LNGTH; BSNW & BSST: WIDTH, GAGE, TEETHPERIN

Aux. Code Format and Dimension Definition
Aux. I.D.: [BSH, BSNW, BSST] D/1: Width D/2: Teeth per Inch

EXAMPLES: BAND SAW BLADES

Tool No.	Description	Aux.	D/1	D/2
BSH1/2.028	1/2W,.028G,4TH,LNGTH,HONEYCOMBBS	BSH	0.5	4
BSH1/2.032	1/2W,.028G,4TH,LNGTH,HONEYCOMBBS	BSH	0.5	4
BSNW3/16.0206	3/16W,.020G,6T,NRRWWDWRKNG BANDSAW	BSNW	0.1875	6
BSNW3/16.0284	3/16W,.028G,4T,NRRWWDWRKNG BANDSAW	BSNW	0.1875	4
BSNW3/16.0286	3/16W,.028G,6T,NRRWWDWRKNG BANDSAW	BSNW	0.1875	6
BSST3/16.0284	3/16W,.028G,4T,SPRNGTMPRD BANDSAW	BSST	0.1875	4
BSST3/16.0286	3/16W,.028G,6T,SPRNGTMPRD BANDSAW	BSST	0.1875	6

TOOL NUMBER CONSTRUCTION GUIDE — BAND SAW BLADES

BST – TUNGSTEN CARBIDE RAKER & STRAIGHT SET

TOOL NO.: BST/WIDTH/SETTYPE/MODIFIER

CLASS	SIZE		MODIFIER
BST	**1-1/4**	**R**	**.05**
BST	WIDTH	SET TYPE	GAGE DIM.
Tungsten Carbide	Fractional	R	Decimal
Band Saw Blade	1-1/4	Raker	.05
			S
			Straight

BST CLASS OF TOOLS: TOOL DESCRIPTION FORMAT
WIDTH, GAGE, PITCH, NOMSET, SETTYPE, TUNGCARB

Aux. Code Format and Dimension Definition
Aux. I.D.: BST D/1: Width D/2: Pitch

EXAMPLES: BAND SAW BLADES

Tool No.	Description	Aux.	D/1	D/2
BST1R	1W,.035G,2-1/2P,.058SET,RKR,TUNGCARB	BST	1	2.5
BST1S	1W,.035G,2-1/2P,.058SET,STR,TUNGCARB	BST	1	2.5
BST1-1/4R	1-1/4W,.035G,2-1/2P,.058SET,RKR,TUNG	BST	1	2.5
BST1-1/4R.05	1-1/4W,.05G,2-1/2P,.075SET,RKR,TUNGC	BST	1	2.5

TOOL NUMBER CONSTRUCTION GUIDE — BAND SAW BLADES

BSV – VTOOTH SINGLE, DOUBLE, & DOUBLE BEVEL DOUBLE EDGE

TOOL NO.: BSV/WIDTH/BEVEL/TEETHPERIN./*EDGE/*MODIFIER
*EDGE: if necessary, specify R, H, or D for Regular, Hard Edge, or Double Edge. * If duplicate exists, add a G for the larger gage diameter.

CLASS		SIZE		MODIFIER
BSV	**3/4**	**D**	**10**	**HG**
BSV Vtooth Band Saw Blade	WIDTH Fractional 3/4	BEVEL S Single D Double	TEETH/IN. 10	EDGE TYPE H - Hard Edge GAGE MOD. G - add to NO. of larger gage dimension

BSV CLASS OF TOOLS: TOOL DESCRIPTION FORMAT
WIDTH, TEETHPERIN., GAGE,EDGE, BEVEL, VTOOTHBANDSAW

Aux. Code Format and Dimension Definition
Aux. I.D.: BSV D/1: Width D/2: Teeth per Inch

EXAMPLES: BAND SAW BLADES

Tool No.	Description	Aux.	D/1	D/2
BSV3/4D10R	3/4W,10T,.020G,REGEDG,DBVL,VTOOTHBS	BSV	0.75	10
BSV3/4D10H	3/4W,10T,.020G,HRDEDG,DBVL,VTOOTHBS	BSV	0.75	10
BSV3/4D10HG	3/4W,10T,.025G,HRDEDG,DBVL,VTOOTHBS	BSV	0.75	10
BSV3/4S10	3/4W,10T,.032G,STDEDG,SBVL,VTOOTHBS	BSV	0.75	10

TOOL NUMBER CONSTRUCTION GUIDE — POWER HACK SAW BLADES

HS – POWER HACKSAW BLADES

TOOL NO.: HS/LENGTH/-WIDTH/-TEETHPERIN.

CLASS	SIZE
HS	**12 - 1 - 10**
HS Power Hack Saw Blade	LENGTH -WIDTH -TEETH/IN. Fractional 12 -1 -10

HS CLASS OF TOOLS: TOOL DESCRIPTION FORMAT
LNGTH, WIDTH, GAGE, TEETHPERIN., POWERHACKSAWBLADE

Aux. Code Format and Dimension Definition
Aux. I.D.: HS D/1: Length D/2: Teeth per Inch

EXAMPLES: HACK SAWS

Tool No.	Description	Aux.	D/1	D/2
HS12-1-10	12L,1W,.05GAGE,10TEETH,PWRHCKSWBLDE	HS	12	10
HS12-1-14	12L,1W,.05GAGE,14TEETH,PWRHCKSWBLDE	HS	12	14

TOOL NUMBER CONSTRUCTION GUIDE — SEGMENTED BAND SAW FILE BLADES

BF – SEGMENTED BAND SAW FILE BLADES

TOOL NO.: BF/WIDTH/TYPE/CUT/TEETHPERIN./-BANDLENGTH

CLASS	SIZE			MODIFIER	
BF	**1/2**	**F**	**C**	**10**	**-75**
BF	WIDTH	TYPE	CUT	TEETH/IN.	LENGTH
Segmented File	Fraction	F	C	10	Inch
Band Saw Blade	1/2	Flat	Coarse		-75
		O	M		
		Oval	Medium		
		H			
		Half Round			

BF CLASS OF TOOLS: TOOL DESCRIPTION FORMAT
WIDTH, TYPE, CUT, TEETHPERIN., LNGTHOF BAND, BANDFILE

Aux. Code Format and Dimension Definition
Aux. I.D.: BF D/1: Width D/2: Teeth

EXAMPLES: BAND FILES

Tool No.	Description	Aux.	D/1	D/2
BF1/2FC10-75	1/2W,FLAT,COARSE,10T,75L,BANDFILE	BF	0.5	10
BF1/2OM14-75	1/2W,OVAL,MEDIUM,14T,75L,BANDFILE	BF	0.5	14
BF1/2FM14-75	1/2W,FLAT,MEDIUM,14T,75L,BANDFILE	BF	0.5	14

VIII. Carbide and Insert Tooling

Carbides and insert tooling (and toolholders) are the most widely used cutting tools for single point machining of cast iron and steel, but are not limited to use on other workpiece materials. They are also used in multipoint tools, such as facemills and indexable insert drills. Carbide is widely brazed on lathe-tool shanks, boring bars, reamers, milling cutters, and circular saws, to name a few. Some tools are available in solid carbide form as well.

HSSs and carbides are not classified in the same way. Grading in the U.S. is predominantly according to the C-classification system; while in Europe and Japan, the ISO system prevails. It is important to note that different carbides that fall into the same grade under either the C-classification or the ISO system are not necessarily equivalent to each other. It is important to refer to the carbide manufacturer's specifications or to use their grading system.

Constructing Tool Numbers and Aux. I.D.'s for CARBIDE TOOLS

Considerations Used for General Tool Number Construction

TYPE	MODIFIER	MODIFIER	MODIFIER
OF			
TOOLS	INDUSTRY		
	CODES		
CARBIDE	STYLES	GRADES	GROUND
BLANKS			UNGROUND
CARBIDE	STANDARD	TURNING BORING	STANDARD TOOL BITS
TIPPED	CODED	FACING THREADING	
	STANDARD TOOLS	ENDMILLS,DRILLS,ETC.	SOLID CARBIDE
INSERT	ITCS CODING	ENDMILLS, SHELL MILLS	STRAIGHT SHANKED
CUTTERS		FACE MILLS,	TAPER SHANKED
		MILLING CUTTERS	ARBOR MOUNTED
INSERTS	ANSI CODES	DIFFERENT GEOMETRIES	TURNING,BORING
	ISO GRADING	& GRADES	THREADING,MILLING
INSERT	ANSI CODES	TURNING,FACING	THREADING
TOOLHOLDER	FRACTIONAL	BORING,MILLING	CARTRIDGES
		INCLUDED ANGLES	CARBIDE

CARBIDE BLANKS, INSERTS MILLING CUTTERS, CARBIDE TIPPED TOOLS

Tool No. Begins	General Description	TLC Aux. I.D.
CBLK	Carbide Blanks -Industry Standard Styles	BLNK
CTB	Carbide Tipped Tool Bits- Industry Standard Styles	CTBx
IEM	Insert Endmill Cutters	CIEM
IMPE	Multiphase Insert Endmill Cutters	
IFM	Insert Face Mill Cutters	CIFM
IMPF	Multiphase Insert Face Mill Cutters	
IHM	Insert Heavy Duty End & Shell Mill Cutters	CIEM
IMC	Insert Milling Cutters	CIM
	Inserts - ANSI Standard Coding	Ixxx
	Insert Toolholder - ANSI Standard Coding	THxx
	Standard Tools - Endmills, Reamers, etc. see tool class with CT modifier for CarbideTipped and SC for Solid Carbide	

TOOL NUMBER CONSTRUCTION GUIDE — CARBIDE BLANKS

CBLK - CARBIDE BLANKS

TOOL NO.: CBLK/STYLENO./ISOGRADE

CLASS	SIZE	MODIFIER
CBLK	**1510**	**K12**
CBLK	Style	ISO Grade
Carbide Blank	1510	K12

CBLK CLASS OF TOOLS: TOOL DESCRIPTION FORMAT
THK%WIDTH%LENGTH,GRADE,CARBIDE BLANK

Aux. Code Format and Dimension Definition

Aux. I.D.: BLNK D/1: Style Number D/2:

or D/1: Thickness D/2: Width

EXAMPLES: CARBIDE BLANKS

Tool No.	Description	Aux.	D/1	D/2
CBLK1510K12	3/8x5/8x1,ISOK12, CARBIDE BLANK	BLNK	1510	
CBLK2330K12	3/16x1/2x3/4,ISOK12 CARBIDE BLANK	BLNK	2330	
CBLK3460K12	5/16x5/8x1 ,ISOK12 CARBIDE BLANK	BLNK	3460	
CBLK4460K12	5/16x5/8x1 ,ISOK12 CARBIDE BLANK	BLNK	4460	
CBLK5340K12	3/16x3/4x3/4,ISOK12 CARBIDE BLANK	BLNK	5340	
CBLK6340K12	3/16x3/4x3/4,ISOK12 CARBIDE BLANK	BLNK	6340	
CBLKRT11K12	3/64x3/32x7/8,1/4R,K12, CARBIDE BLNK	BLNK	0.046	0.094
CBLKCT140K12	7/8D,1-3/8L, ISOK12 CARBIDE BLANK	BLNK	0.875	1.375
CBLKSTB824K12	1/4x3/4x6, ISOK12 CARBIDE BLANK	BLNK	0.25	0.75
CBLKSTB616BK12	3/16x1/2x2-1/2,ISOK12,CARBIDE BLANK	BLNK	0.1875	0.5

TOOL NUMBER CONSTRUCTION GUIDE — CARBIDE TIPPED TOOL BITS

CTB – CARBIDE TIPPED TOOL BITS

TOOL NO.: CTB/STYLENO./ISOGRADE

CLASS	SIZE	MODIFIER
CTB	**AL16**	**K12**
CTB Carbide Tipped Tool Bit	Style AL-16	ISO Grade K12

CBT Class of Tools: Tool Description Format
SHANK%LENGTH, STYLE, GRADE, TOOL BIT

Aux. Code Format and Dimension Definition

Aux. I.D.: CTBx D/1: Shank D/2: Length

x = A-Turning, B-Lead turning, C-Square nose, D-Pointed nose, E-Threading, F-End cutting, G-Side cutting, O-Cutoff, V-Grooving, T-Boring

EXAMPLES: CARBIDE TIPPED TOOL BITS

Tool No.	Description	Aux.	D/1	D/2
CTBAR4K12	1/4SHNK,2L,ISOK12,0LEADTURN, TOOLBIT	CTBA	0.25	2.0
CTBAL4K12	1/4SHNK,2L,ISOK12,0LEADTURN,LHTOOLBIT	CTBA	0.25	2.0
CTBAL16K12	1SHNK,6L ,ISOK12,0LEADTURN,LHTOOLBIT	CTBA	1.0	6.0
CTBBR4K12	1/4SHNK,2L,ISOK12,15LEADTRN, TOOLBIT	CTBB	0.25	2.0
CTBBL4K12	1/4SHNK,2L,ISOK12,15LEADTRN,LHTOOLBIT	CTBB	0.25	2.0
CTBC4K12	1/4SHNK,2L,ISOK12,SQNOSE, TOOLBIT	CTBC	0.25	2.0
CTBD4K12	1/4SHNK,2L,ISOK12,64PTDNOSE, TOOLBIT	CTBD	0.25	2.0
CTBE5K12	5/16SHNK,2-1/4L,K12,THREADING,TOOLBIT	CTBE	0.3125	2.25
CTBCOR443K12	1/2x1x5,ISOK12,CUTOFF TOOLBIT	CTBO	0.5	5.0
CTBFR12K12	3/4SHNK,4.5L,ISOK12,ENDCUTTNG,TOOLBIT	CTBF	0.75	4.5
CTBFL12K12	3/4SHNK,4.5L,K12,ENDCUTTNG,LH TOOLBIT	CTBF	0.75	4.5
CTBGR12K12	3/4SHNK,4.5L,ISOK12,SIDECUTTNGTOOLBIT	CTBG	0.75	4.5
CTBGL12K12	3/4SHNK,4.5L,K12,SIDECUTTNG,LHTOOLBIT	CTBG	0.75	4.5
CTBVG21K12	5/8x1x6,ISOK12,GROOVE TOOLBIT	CTBV	0.625	6.0
CTBTSA6K12	3/8SHNK,1-3/4L,K12,BORING TOOLBIT	CTBT	0.375	1.75
CTBTRC6K12	3/8D,1-3/4L,K12,60 RND BORING TOOLBIT	CTBT	0.375	1.75
CTBTRE6K12	3/8D,1-3/4L,K12,45 RND BORING TOOLBIT	CTBT	0.375	1.75
CTBTSC6K12	3/8SHNK,1-3/4L,K12,60 BORING TOOLBIT	CTBT	0.375	1.75
CTBTSE6K12	3/8SHNK,1-3/4L,K12,45 BORING TOOLBIT	CTBT	0.375	1.75
CTBTSB6K12	3/8SHNK,1-3/4L,K12,90 BORING TOOLBIT	CTBT	0.375	1.75

TOOL NUMBER CONSTRUCTION GUIDE — INSERT END, FACE & SHELL MILLS

IEM, IFM, IHM, IMPE, IMPF INSERTED ENDMILL, FACE MILL, SHELL MILL HOLDERS

TOOL NO.: Class Code/DIA/RAKE or LEAD/#SLOTS/SHANKSTYLE

CLASS	SIZE	RAKE -LEAD	CUT STATIONS	SHANK -BORE
IEM	**1-1/2**	**0**	**2F**	**S**
IEM	Dia.	Rake	# of Flutes	S
Insert	1-1/2	0	2F	Strght Shank
Endmill		Neutral		
IMPE		N		R8
Multiphase		Negative		R8 Shank
		P		
		Positive		#40
IFM				40NMTB
Insert		Lead Angle	# of Slots	
Face Mill		3		#50
IMPF				50NMTB
Multiphase				
IHM				
Insert Heavy				B
Duty End or				Bore no
Shell Mill				shank

INSERT MILL CLASS OF TOOLS: TOOL DESCRIPTION FORMAT
DIA, #SLOTS, RAKE/LEAD, SHANK, TYPEOFCUTTER

Aux. Code Format and Dimension Definition

Aux. I.D.: [CIEM, CIFM]   D/1: Diameter   D/2: Shank Diameter

EXAMPLES: INSERT MILLS

Tool No.	Description	Aux.	D/1	D/2
IEM1-1/202FS	1-1/2D,2F,0RAKE,1STRSHNK, ENDMILLCUTT	CIEM	1.500	1.000
IEM1-1/2P2FS	1-1/2D,2F,POSRK,1STRSHNK, ENDMILLCUTT	CIEM	1.500	1.000
IEM2-1/2N4FS	2-1/2D,4F,NEGRK,1.25SHNK, ENDMILLCUTT	CIEM	2.500	1.250
IFM2-1/2153FR8	2-1/2D,3F,15LEAD,R8SHNK, FACEMILLCUTT	CIFM	2.500	0.9490
IFM2-1/2153FS	2-1/2D,3F,15LEAD,1.25SHK,FACEMILLCUTT	CIFM	2.500	1.250
IMPE1-1/29F#40	1-1/2D,9F,MULTI,#40NMTB,ENDMILLCUTTER	CIEM	1.500	40
IMPF406FB	4D,6F,MULTI,1.5BORE,0LEAD,FACEMILLCUT	CIFM	4.000	1.5
IMPF4306FB	4D,6F,MULTI,1.5BORE,30LEAD,FACEMILLCT	CIFM	4.000	1.5
IHM338FS	3D,8F,3LEAD,1.25SHNK,MULTIHD,ENDMILLC	CIEM	3.000	1.250
IHM3158FS	3D,8F,15LEAD,1.25SHNK,MULTIHD,ENDMILL	CIEM	3.000	1.250
IHM4158FB	4D,8F,15LEAD,1.5BORE,MULTIHD,SHELLMIL	CIEM	4.000	1.500

Constructing Tool Numbers and Aux. I.D.'s for CUTOFF BLADES

Considerations Used for General Tool Number Construction

TYPE			
OF	FRACTIONAL		
BLADE			
			M-2
			M-42
STANDARD	WIDTH	VARIOUS STYLES	
T-SHAPED	IN 1/16THS	B&S	T-15
TAPERED		ARMSTRONG	
RECTANGLE	HEIGHT	PRATT&WHITNEY	
ROUND	IN 1/16THS	EMPIRE	
DBL BEVEL		LUERS	
SQUARE	LENGTH		
	IN INCHES		

CUTOFF BLADES

Tool No. Begins	General Description	TLC Aux. I.D.
COB	Cutoff Blades	COB

TOOL NUMBER CONSTRUCTION GUIDE — CUTOFF BLADES

COB – CUTOFF BLADES

TOOL NO.: COB/TYPE/W/H/L/MODIFIER

CLASS	TYPE	SIZE	MODIFIER
COB	**RF**	**8 6 3**	**T15**
COB	RF	WIDTH,HEIGHT,LNGTH	Mat'l
Cutoff Blade	Rectangle Fin.Grnd	In 32's In 16's In "s	M-2
	PT	8/32 6/16 3	M-42
	Parallel T-Shaped	1/4 3/8 3	T-15
	TG		
	Tapered Gen.Pur.		L
	R		Left Hand
	Rectangular		
	RD		
	Round		
	DB		
	Double Bevel		
	BS		
	B&S Style		
	AS		
	Armstrong Style		
	PW		
	Pratt&Whitney Style		
	E or L		
	Empire or Luers		
	SQ		
	Square		
	SB		
	Square Beveled 15 deg.		
	Omit		
	if standard		

COB CLASS OF TOOLS: TOOL DESCRIPTION FORMAT
SIZE, MODIFIER, TYPE CUTOFF BLADE

Aux. Code Format and Dimension Definition

Aux. I.D.: COB D/1: Width D/2: Height

EXAMPLES: CUTOFF BLADES

Tool No.	Description	Aux.	D/1	D/2
COBPT284.5	1/16x1/2x4.5,PARALLEL T-SHAPE CUTOFF	COB	0.0625	0.5
COBPT484.5	1/8x1/2x4.5,PARALLEL T-SHAPE CUTOFF	COB	0.125	0.5
COBPT4115	1/8x11/16x5,PARALLEL T-SHAPE CUTOFF	COB	0.125	0.6875
COB384.5L	3/32x1/2x4.5,STRAIGHT,LHAND CUTOFF	COB	0.094	0.5
COBTG284.5	1/16x1/2x4.5,TAPERED,CUTOFF BLADE	COB	0.0625	0.5
COBTG2115	1/16x11/16x5,TAPERED,CUTOFF BLADE	COB	0.0625	0.6875
COBR464	1/8x3/8x4,RECTANGULAR, CUTOFF BLADE	COB	0.125	0.375
COBRD64	3/16Dx4,ROUND, CUTOFF BLADE	COB	0.1875	
COBDB384.5	3/32x1/2x4.5,DBL BEVEL,CUTOFF BLADE	COB	0.094	0.5
COBBS384.5	3/32x1/2x4.5,B&S, CUTOFF BLADE	COB	0.094	0.5
COBAS4126	1/8x.715x6,ARMSTRONG,CUTOFF BLADE	COB	0.125	0.715
COBPW284.5	1/16x1/2x4.5,PRATT&WHITNEY CUTOFF	COB	0.0625	0.5
COBE3115	3/32x.676x5,EMPIRE,CUTOFF BLADE	COB	0.094	0.676
COBL3115	3/32x.676x5,LUERS ,CUTOFF BLADE	COB	0.094	0.676
COBSQ632.5	3/16SQx2.5,SQUARE CUTOFF BLADE	COB	0.1875	0.1875
COBSB632.5M2	3/16SQx2.5,M-2,SQ.BEVEL,CUTOFF BLADE	COB	0.1875	0.1875
COBSB632.5T15	3/16SQx2.5,T-15,SQ.BEVEL,CUTOFFBLADE	COB	0.1875	0.1875
COBSB632.5M42	3/16SQx2.5,M-42,SQ.BEVEL,CUTOFFBLADE	COB	0.1875	0.1875
COBRF863M2	1/4x3/8x3,M-2,RECT,FIN.GRND,CUTOFF	COB	0.25	0.375
COBRF863T15	1/4x3/8x3,T-15,RECT.FIN.GRND,CUTOFF	COB	0.25	0.375

Constructing Tool Numbers and Aux. I.D.'s for INSERTS AND INSERT TOOL HOLDERS

Basically, ITCS recommends use of the ANSI or ISO standard for inserts and insert tool holders. The various manufacturers of these tools all use one of these standards with modifications for the features they individually add to the product.

The samples shown by the following are for your reference. You should use the numbers provided for these products by their manufacturer or the identification best known by your shop personnel. The most confusing part will be putting a grade on inserts. You could construct an insert number of each manufacturer of the insert that you have with the manufacturer's grade specified, or you could pick a standard such as ISO and use it throughout. This would mix manufacturers of the same insert in your storage bins. Overall, the storage and numbering of inserts is going to take some planning on your part. Many manufacturers of inserts offer a consulting service to eliminate duplication and to select the inserts and grades best suited to your needs.

INSERTS

ANSI Standard

N	M	G	4	3	4	E			
SHAPE	RELIEF	TOLER	TYPE	I.C.	THKNESS	CUT	PNT	EDGE	DIRECTN
+	+	+	+	+	+	+	+	+	+
N	U	N	12	04	08	E	N		

ISO Standard
Aux. I.D.: IN (SHAPE CODE) D/1: Thickness D/2: Radius

INSERTS TOOL HOLDERS

ANSI Standard

M	T	G	N	R	16	4	D	
HOLDNG	IC. SHPE	STYLE	RAKE	HAND	SHANK	I.C. SIZE	SHNK QUAL	MODS

HOLD	IC. SHPE	STYLE	CLEAR	HAND	HGHT	WDTH	LNGTH	IC. SZE	SHNK	MOD
+	+	+	+	+	+	+	+	+	+	+
P	S	S	N	R	3	2	3	2 -	19	Q -

ISO Standard
Aux. I.D.: IH (INSERT SHAPE) (RELIEF ANGLE)
D/1: Width D/2: Length

IX. Burrs, Files, and Abrasives

Filing and burring is a method of smoothing a surface with a minimum of material removal. Files and burrs present numerous cutting edges or ridges to the workpiece to minimize the time and labor required to remove the material with a single cutting edge and work stroke.

The most common tools for filing and burring are hand files of Swiss and American patterns, rotary files, rifflers, and burrs. Files and burrs can be supplied in HSS, solid carbide or (burrs) with steel shanks and carbide heads. These are generally hand tools, but can also be used in powered machines and portable power tools.

These tools are available in a variety of head shapes, patterns, and sizes. Rifflers, deburring, and chamfer cutters have also been included because of their finishing operation associated function.

Constructing Tool Numbers and Aux. I.D.'s for BURRS & FILES

Considerations used for General Tool Number Construction

TYPE OF	STYLE	MODIFIER	MODIFIER
BURRS			
OR			
FILES			
MIDGET &		VARIOUS CUTS	
JUNIOR	GEOMETRIC		HSS
MILLS	SHAPES		CARBIDE
	STOCK NO.S		
DEBURRING			
&	FRACTIONAL	INSIDE/OUTSIDE	
CHAMFERING		BALL SEAT	
CUTTERS			
FILES	AMERICAN	STYLES	
	SWISS	SHAPES	
	NEEDLE	CUTS	
	PARALLEL		
RIFFLERS	MODEL NO.	SHAPES	
		CUTS	

BURRS, DEBURRING & CHAMFERING CUTTERS, FILES, RIFFLERS

Tool No. Begins	General Description	TLC Aux. I.D.
	Burrs, Midget & Junior Mills - 1/8, 3/16, 1/4, 3/8 Diameter Shanks - HHS & Carbide Heads w/Steel Shanks, Solid Carbide	BURR
ODC	Outside Deburring Cutters	COD
IDC	Inside Deburring Cutters	CDI
IODC	Inside/Outside Deburring Cutters	CDIO
OCC	Outside Chamfer Cutters	CCFO
ATCC	Adjustable Tube Chamfer Cutters	CCFA
BSR	Ball Seat Reamers	RMS
FA	American Files	FIA
FS	Swiss Files	FIS
FN	Needle Files	FIN
FP	Parallel Files	FIP
DSR	Die Sinkers Rifflers	RIF

Notes:

Most solid carbide burrs come in 1/8, 3/16, and 1/4 diameters. 1/8 diameter tools are 1-1/2 OAL, 3/16 and 1/4 are 2 OAL.

TOOL NUMBER CONSTRUCTION GUIDE — BURRS

TOOL NO.: HEAD DIA or RAD./STYLE/CUT/MODIFIERS

HEAD SIZE	CLASS	CUT TYPE	MODIFIERS		
1/8	**CY**	**MS**	**EC**	**25**	**CB**
FRACTIONAL	Head Style	Cut Style	EC	Shank Dia.	Head
1/8	CY	MS	End Cut	1/8 = 12	Mat'l
Head Dia.	Cylinder	Medium Stand	L	3/16 = 18	C
	BN	MM	Long	1/4 = 25	Carbide
use Radii	Ball Nose	Medium Master		3/8 = 38	S
for	B	MC		Omit if	Solid
Concave	Ball	Medium Chpbrk		= to head	Carbide
Concave-Convex	O	MD		or Solid	
Multiple Concave	Oval	Med. Diamond		Carbide	Omit
	TR	CS			for
	Tree Radius	Coarse Standard			HSS
	TP	CM			
	Tree Point	Coarse Master			
	FL	CC			
	Flame	Coarse Chipbrkr			
	Cxx	CD			
	Cone Inc Angle	Coarse Diamond			
	C	FS			
	Cone	Fine Standard			
	ICxx	FM			
	Inverted Cone	Fine Master			
	CC	FC			
	Concave-Convex	Fine Chipbrkr			
	CV	FD			
	ConcaVe	Fine Diamond			
	MC				
	Multiple Concave				
	P				
	Pear				

BURR CLASS OF TOOLS: TOOL DESCRIPTION FORMAT
DIA. Or RAD., STYLE, FLUTELNGTH, MAT'L, MOD.S, BURR

Aux. Code Format and Dimension Definition

Aux. I.D.: BURR D/1: Diameter D/2: Shank Diameter

EXAMPLES: BURRS

Tool No.	Description	Aux.	D/1	D/2
1/8CYMSEC25	1/8D,CYLNDR,1/2FL,1/4SHNK, BURR	BURR	0.1255	0.25
3/4CCMS25	3/4RAD,CNCV-CNVX,1FL,1/4HEADDIA,BUR	BURR	0.75	0.25
1CCMS25	1RAD,CNCV-CNVX,1-1/8FL,5/16HEAD,BURR	BURR	1.00	0.25
1/16MCMS25	1/16RAD,MLTPLE-CNCV,1/8FL,1/4HEAD,BR	BURR	0.0625	0.25
3/32CVMS25	3/32RAD,CNCV,1/8FL,1/4HEAD, BURR	BURR	0.09375	0.25
5/16CVMS25	5/16RAD,CNCV,3/8FL,1/2HEAD BURR	BURR	0.3125	0.25
5/16PMS25	5/16D,PEAR,3/8FL, BURR	BURR	0.3125	0.25
1/2PMS25	1/2D,PEAR,5/8FL, BURR	BURR	0.5000	0.25
1/4IC10MS12C	1/4D,INVCONE10DEG,1/4FL,CARBIDE,BURR	BURR	0.2500	0.125
1/4C22MS12C	1/4D,22DEGCONE,1/2FL,CARBIDE, BURR	BURR	0.2500	0.125
1/4TPMS12C	1/4D,TREEPNT,1/2FL,CARBIDE, BURR	BURR	0.2500	0.125
1/4TRMS12C	1/4D,TREERAD,1/2FL,CARBIDE, BURR	BURR	0.2500	0.125
1/4OMS12C	1/4D,OVAL,3/8FL,CARBIDE, BURR	BURR	0.2500	0.125
1/4BMS12C	1/4D,BALL,1/4FL,CARBIDE, BURR	BURR	0.2500	0.125
1/4BNMS12C	1/4D,BALLNOSE,1/2FL,CARBIDE, BURR	BURR	0.2500	0.125
1/4CYMS12C	1/4D,CYLNDR,1/2FL,CARBIDE, BURR	BURR	0.2500	0.125
1/4CYMS	1/4D,CYLNDR,3/4FL, BURR	BURR	0.2500	0.25
1/4BNMS	1/4D,BALLNOSE,3/4FL, BURR	BURR	0.2500	0.25
1/4BMS	1/4D,BALL,1/4FL, BURR	BURR	0.2500	0.25
1/4OMS	1/4D,OVAL,5/8FL, BURR	BURR	0.2500	0.25
1/4TRMS	1/4D,TREERAD,3/4FL, BURR	BURR	0.2500	0.25
1/4C14MS	1/4D,CONERAD14DEG,3/4FL, BURR	BURR	0.2500	0.25
3/8C60MS25C	3/8D,CONE60DEG,5/16FL,CARBIDE, BURR	BURR	0.375	0.25
3/8CMS25C	3/8D,CONE,5/8FL,CARBIDE, BURR	BURR	0.375	0.25
3/8ICMS25C	3/8D,INVTCONE,3/8FL,CARBIDE, BURR	BURR	0.375	0.25
3/8C90MS25C	3/8D,CONE90DEG,3/16FL,CARBIDE, BURR	BURR	0.375	0.25
3/8OMS25C	3/8D,OVAL,5/8FL,CARBIDE, BURR	BURR	0.375	0.25
3/8TRMS25C	3/8D,TREERAD,3/4FL,CARBIDE, BURR	BURR	0.375	0.25
5/16CYMS25C	5/16D,CYLNDR,3/4FL,CARBIDE, BURR	BURR	0.3125	0.25
5/16BNMS25C	5/16D,BALLNOSE,3/4FL,CARBIDE, BURR	BURR	0.3125	0.25
5/16BMS25C	5/16D,BALL,1/4FL,CARBIDE, BURR	BURR	0.3125	0.25
5/16TPMS25C	5/16D,TREEPOINT,3/4FL,CARBIDE, BURR	BURR	0.3125	0.25
5/16FLMS25C	5/16D,FLAME,3/4FL,CARBIDE, BURR	BURR	0.3125	0.25
5/16C14MS25C	5/16D,CONERAD14DEG,7/8FL,CARBIDEBUR	BURR	0.3125	0.25
1/8CYMSS	1/8D,CYLNDR,9/16FL,SOLIDCARBIDE,BURR	BURR	0.125	0.125
1/8TPMSS	1/8D,TREEPNT,1/2FL,SOLIDCARBIDE,BURR	BURR	0.125	0.125
1/8BNMSS	1/8D,BALLNOSE,9/16FL,SOLIDCARB, BURR	BURR	0.125	0.125
1/8C60MSS	1/8D,CONE 60 DEG, SOLIDCARBIDE,BURR	BURR	0.125	0.125
1/8BMSS	1/8D,BALL ,1/8 FL,SOLIDCARBIDE,BURR	BURR	0.125	0.125
1/8C90MSS	1/8D,CONE 90 DEG, SOLIDCARBIDE,BURR	BURR	0.125	0.125
1/8OMSS	1/8D,OVAL, 7/32FL,SOLIDCARBIDE,BURR	BURR	0.125	0.125
1/8C8MSS	1/8D,CONE8DEG,1/2FL,SOLIDCARB, BURR	BURR	0.125	0.125
1/8TRMSS	1/8D,TREERAD,1/2FL,SOLIDCARBIDE,BURR	BURR	0.125	0.125
1/8IC10MSS	1/8D,INVRTCONE10DEG,3/16FL,SOLIDCARB	BURR	0.125	0.125

Note: You may wish to simplify your numbering of burrs by using the style number of the particular burrs you currently stock.

TOOL NUMBER CONSTRUCTION GUIDE — DEBURRING AND CHAMFER CUTTERS

ODC, IDC, IODC, OOC, ATCC, BSR — OUTSIDE, INSIDE, INSIDE/OUTSIDE DEBURRING CUTTERS & OUTSIDE , ADJUSTABLE TUBE CHAMFER CUTTERS AND BALL SEAT REAMERS

TOOL NO.: Class Code/DIAMETER/ANGLE/MODIFIER

CLASS	SIZE		MODIFIER
IDC	**1/2**	**60**	**CB**
ODC	Diameter	Angle	Material
Outside	Rod/End	Included	HSS
Deburring	1/2	60	Standard
IDC			Large end
Inside			CB
Deburring			Carbide
IODC	O.D.		
Inside/Outside			
Chamfer			
OCC	Bar Dia.		
Outside			
Chamfer			
ATCC	Tube O.D.		
Adjustable			
Tube Chamfer			
BSR	Dia.		
Ball Seat			
Reamer			

DEBURRING CLASS OF TOOLS: TOOL DESCRIPTION FORMAT
DIAMETER,BODY,ANGLE,CHAMFER CUTTER

Aux. Code Format and Dimension Definition

AUX I.D.:	CDO	D/1: ROD DIA.	D/2: ANGLE
	CDI	LARGE END	INC. ANGLE
	CDIO	TUBE O.D.	TUBE I.D.
	CCFO	MAJ. BAR DIA	ANGLE
	CCFA	TUBE O.D.	PLUG
	RBS	DIA	SHANK DIA

EXAMPLES: DEBURRING & CHAMFER TOOLS

Tool No.	Description	Aux.	D/1	D/2
ODC3/1630	3/16ROD,1/2BODY,30DEG,OUTSIDE DEBUR	CDO	0.1875	30
IDC1/460	1/4LARGE,0SMALL,60DEG, INSIDE DEBUR	CDI	0.25	60
IODC1/8	1/8O.D.,1/16I.D.,IN/OUT DEBURRING	CDIO	0.125	0.0625
OCC1/230	1/2BARDIA,7/89BODY,30DEG, OUT CHAMF	CCFO	0.5	30
ATCC3/16	3/16O.D.,.152PLUG,ADJ.TUBE CHAMFER	CCFA	0.1875	0.152
BSR1/4	1/4D,3/16SHNK, BALL SEAT REAMER	RBS	0.25	0.1875

TOOL NUMBER CONSTRUCTION GUIDE — FILES AND DIE SINKERS RIFFLERS

TOOL NO.: Class Code/LENGTH or STOCKNO./SHAPE/TYPEOFCUT

CLASS	SIZE	SHAPE	TYPE OF CUT
FA	**14**	**M**	**B**
FA	14	M	B
American	Length	Mill Style Shape	Bastard Cut
Style File	in		
	Inches	CUTS TABLE	
FS			
Swiss	FOR		FOR AMERICAN:
Style File			
	SHAPES		B - Bastard
FN			2C - Second Cut
Needle	REFER		SC - Smooth Cut
Style File			
	TO		
FP			SWISS:
Parallel	TABLE		Input Cut Number
Style File			00, 0, 1, 2, 3, 4,
			5, 6, 8
DSR	STOCK NO.		
Die Sinkers			
Riffler			

SHAPES TABLE

CODE	SHAPE DESCRIPTION	AVAILABLE PATTERNS
M -	Mill	American
M1R -	Mill One Round Edge	American
M2R -	Mill Two Round Edge	American
TR -	Taper Regular	American
TS -	Taper Slim	American
TX -	Taper Xtra Slim	American
TDX -	Taper Double Xtra Slim	American
P -	Pillar	American, Swiss, Parallel
PN -	Pillar Narrow	Swiss
PNX -	Pillar Xtra Narrow	Swiss
W -	Warding	American, Swiss, Needle
F -	Flat	American
FL -	Flat Long Angle	American
H -	Hand	American, Swiss
HR -	Half Round	American, Swiss, Needle, Parallel
HRS -	Half Round Slim	Swiss
SQ -	Square	American, Swiss, Needle, Parallel
3SQ -	Three Square	American, Swiss, Needle, Parallel
R -	Round	American, Swiss, Needle, Parallel
K -	Knife	American, Swiss, Needle, Parallel
B -	Barrette	Swiss, Needle
E -	Equalling	Swiss, Needle, Parallel
X -	Crossing	Swiss, Needle, Parallel
J -	Joint	Needle
MRK -	Marking	Needle
S -	Slitting	Needle
O -	Oval	Parallel
CR -	Crochet	Parallel
PP -	Pippin	Parallel
L -	Lozenge	Parallel
C -	Cant	Parallel

FA, FS, FN, FP, DSR — AMERICAN, SWISS, NEEDLE, PARALLEL FILES & DIE SINKER'S RIFFLERS CLASS OF TOOLS: TOOL DESCRIPTION FORMATS
LENGTH, APPROX.SIZE, TYPEOFCUT, SHAPE, FILE

Aux. Code Format and Dimension Definition
Aux. I.D.: [FIA, FISW, FIND, FIPA] D/1: Length D/2: Riff, Stock No.

EXAMPLES: FILES

Tool No.	Description	Aux.	D/1	D/2
FA4MB	4",7/16x5/64,BASTARD,MILL,AMER.FILE	FIA	4	
FA4M2C	4",7/16x5/64,2NDCUT, MILL,AMER.FILE	FIA	4	
FA6MSC	6",5/8x7/64,SMOOTHCT,MILL,AMER.FILE	FIA	6	
FA6M1RSC	6",5/8x7/64,SMOOTHCT,MILL1RND .FILE	FIA	6	
FA8M2RSC	8",25/32x1/8,SMOOTH,MILL2RND, .FILE	FIA	8	
FA6TR	6",15/32",SINGLECUT,TAPER AMER.FILE	FIA	6	
FA4TS	4",15/64",SINGCUT,SLIMTAPER,AM.FILE	FIA	4	
FA4TX	4",3/16",1CUT,XTRASLIMTAPER,AM.FILE	FIA	4	
FA4TDX	4",5/32",1CUT,DBLXTRASLMTAPER FILE	FIA	4	
FA6PB	6",7/16x7/32,BASTARD,PILLAR,AM.FILE	FIA	6	
FA6P2C	6",7/16x7/32,2ND CUT,PILLAR,AM.FILE	FIA	6	
FA6PSC	6",7/16x7/32,SMTHCUT,PILLAR,AM.FILE	FIA	6	
FA4WB	4",15/32x3/64,BASTARD,WARDING FILE	FIA	4	
FA4W2C	4",15/32x3/64,2NDCUT, WARDING FILE	FIA	4	
FA4WSC	4",15/32x3/64,SMTHCUT,WARDING FILE	FIA	4	
FA4FB	4",15/32x3/32,BASTARD,FLAT,AMERFILE	FIA	4	
FA4F2C	4",15/32x3/64,2NDCUT FLAT,AMERFILE	FIA	4	
FA4FSC	4",15/32x3/64,SMTHCUT,FLAT,AMERFILE	FIA	4	
FA6HB	6",5/8x5/32,BASTARD,DBLCUTHAND FILE	FIA	6	
FA6H2C	6",5/8x5/32,2NDCUT, DBLCUTHAND FILE	FIA	6	
FA6HSC	6",5/8x5/32,SMTHCUT,DBLCUTHAND FILE	FIA	6	
FA4HRB	4",25/64x7/64,BASTARD,1/2RND FILE	FIA	4	
FA4HR2C	4",25/64x7/64,2NDCUT, 1/2RND FILE	FIA	4	
FA4HRSC	4",25/64x7/64,SMTHCUT,1/2RND FILE	FIA	4	
FA4SQB	4",11/64",BASTARD,SQUARE, AMER.FILE	FIA	4	
FA4SQ2C	4",11/64",2NDCUT, SQUARE, AMER.FILE	FIA	4	
FA4SQSC	4",11/64",SMTHCUT,SQUARE, AMER.FILE	FIA	4	
FA4RB	4",9/64",BASTARD,ROUND, AMER.FILE	FIA	4	
FA4R2C	4",9/64",2NDCUT ,ROUND, AMER.FILE	FIA	4	
FA4RSC	4",9/64",SMTHCUT,ROUND, AMER.FILE	FIA	4	
FA63SQB	6",27/64",BASTARD,3SQUARE,AMER.FILE	FIA	6	
FA63SQ2C	6",27/64",2NDCUT ,3SQUARE,AMER.FILE	FIA	6	
FA63SQSC	6",27/64",SMTHCUT,3SQUARE,AMER.FILE	FIA	6	
FA4KB	4",15/32x7/64,BASTARD,KNIFE,AM.FILE	FIA	4	
FA4K2C	4",15/32x7/64,2NDCUT ,KNIFE,AM.FILE	FIA	4	
FA4KSC	4",15/32x7/64,SMTHCUT,KNIFE,AM.FILE	FIA	4	
FA10FL	10",31/32x13/64,FLATLNGANFLATHEFILE	FIA	10	
FA12FL	12",1-11/64x15/64,FLTLNGANGLTHEFILE	FIA	12	
FA14FL	14",1-3/8x1/4,FLATLNGANGLATHE FILE	FIA	14	
FHANDLE#4	#4 FILE HANDLE FOR 12-16" FILES	FHND	12	16
FHANDLE#3	#3 FILE HANDLE FOR 8-12" FILES	FHND	8	12
FHANDLE#2	#2 FILE HANDLE FOR 6" FILES	FHND	6	
FHANDLE#1	#1 FILE HANDLE FOR 3-6" FILES	FHND	3	6
FCH3/4-6C	3/4x6",COARSE,CARBIDE HAND FILE	FCH	0.75	6
FCH3/4-6MF	3/4x6",MEDIUM FINE,CARBIDE HAND FILE	FCH	0.75	6
FCH3/4-3C	3/4x3",COARSE,CARBIDE HAND FILE	FCH	0.75	3
FCH3/4-3CK	3/4x3",COARSE,CARBIDEKNOBHAND FILE	FCH	0.75	3

Notes on American Pattern Files:

1) Mill, flat, long, angle, lathe, and taper files are provided standard as single cut. 2) Pillar, three square, and square files are provided standard as double cut. 3) Warding and flat files are provided standard as double cut on surfaces and single cut on edges. 4) Hand files are provided standard with one edge uncut. 5) Half round files are provided standard with flats double cut and backs of bastards double cut, backs of second cuts longer than 6" double cut, and backs of smooth cut are single cut. 6) Round files 6" and longer bastard cuts and 12" and longer second cuts are provided standard as double cut. All others are single cut. 7) Carbide Hand files are provided in C (Coarse), MC (Medium Coarse), M (Medium), MF (Medium Fine), and F (Fine) cuts.

EXAMPLES: SWISS PATTERN FILES

Tool No.	Description	Aux.	D/1	D/2
FS4B00	4",00CUT,1/2x1/8,BARRETTESWISS FILE	FISW	4	
FS4B4	4", 4CUT,1/2x1/8,BARRETTESWISS FILE	FISW	4	
FS4E00	4",00CUT,23/64x1/16,EQUALLING, FILE	FISW	4	
FS4E4	4", 4CUT,23/64x1/16,EQUALLING, FILE	FISW	4	
FS4H0	4", 4CUT,17/32x1/8,HAND SWISS FILE	FISW	4	
FS4PN00	4",00CUT,3/16x3/32,NARROWPILLARFILE	FISW	4	
FS4P00	4",00CUT,9/32x3/32,PILLAR SWISS FILE	FISW	4	
FS4PNX00	4",00CUT,1/8x5/64,XTRANRRWPILLARFILE	FISW	4	
FS4W00	4",00CUT,23/64x1/32,WARDINGSWISSFILE	FISW	4	
FS4HR00	4",00CUT,5/16x3/32,HALFROUNDSWISSFLE	FISW	4	
FS6HRS00	6",00CUT,5/16x3/32,SLIMHALFROUNDFILE	FISW	6	
FS43SQ00	4",00CUT, 9/32,3 SQUARE SWISS FILE	FISW	4	
FS4SQ00	4",00CUT,5/32,SQUARE SWISS FILE	FISW	4	
FS4K00	4",00CUT,15/32x1/8,KNIFE SWISS FILE	FISW	4	
FS4R00	4",00CUT,5/32,ROUND SWISS FILE	FISW	4	
FS4X00	4",00CUT,15/32x9/64,CROSSING FILE	FISW	4	

Notes on Swiss Pattern Files:

1) The number of teeth per inch of each cut number vary according to the length of cut of the file. 2) Swiss pattern cut no.'s 00, 0, 2 are comparable to American pattern cuts bastard, second cut, smooth cut. There is no equivalent in American patterns to Swiss patterns for cuts numbered 1, 3, 4, 5, 6, and 8. Barrette files are cut on the flat side only.

EXAMPLES: PARALLEL FILES

Tool No.	Description	Aux.	D/1	D/2
FP3/32R00	3/32",00CUT,ROUND PARALLEL MACHFILE	FIPA	0.09375	
FP1/8S00	1/8",00CUT,SQUARE PARALLEL MACHFILE	FIPA	0.125	
FP1/4P00-2	1/4x1/8",00CUT 2SIDES,PILLAR,PARALL	FIPA	0.250	
FP3/16P00	3/16x1/16",00CUT,PILLAR PARALLELFILE	FIPA	0.1875	
FP3/16CR00	3/16x3/32",00CUT,CROCHET,PARALLEL	FIPA	0.1875	
FP1/43SQ00	1/4",00CUT,3SQUARE,PARALLEL MACHFILE	FIPA	0.250	
FP3/16HR00	3/16x7/64,00CUT,HALFROUND,PARALLEL	FIPA	0.1875	
FP1/4X00	1/4x7/64,00CUT,CROSSING,PARALLEL	FIPA	0.250	
FP11/32O00	11/32x9/64,00CUT,OVAL,PARALLEL FILE	FIPA	0.343	
FP15/32K00	15/32x1/8,KNIFE, PARALLEL MACHFILE	FIPA	0.468	
FP19/64PP00	19/64x5/32,00CUT,PIPPIN, PARALLEL	FIPA	0.296	
FP3/8E00	3/8x1/16,00CUT,EQUALLING, PARALLEL	FIPA	0.375	
FP11/32L00	11/32x17/64,00CUT,LOZENGE PARALLEL	FIPA	0.343	
FP3/8C00	3/8x3/16,00CUT,CANT,PARALLEL FILE	FIPA	0.125	

Notes on Swiss Pattern Parallel Machine Files:

For basically all American precision filing machines, and Oliver, Harvey, Butterfly, Cochrane-Bly and similar filing machines. Designed for cut on downward stroke. Length of cut is 6", overall length is 8".

EXAMPLES: NEEDLE FILES

Tool No.	Description	Aux.	D/1	D/2
FN5.5K00	5-1/2",00CUT, KNIFE NEEDLE FILE	FIND	5.5	
FN5.5J00	5-1/2",00CUT, JOINT NEEDLE FILE	FIND	5.5	
FN5.5S00	5-1/2",00CUT, SLITTING NEEDLE FILE	FIND	5.5	
FN5.5X00	5-1/2",00CUT, CROSSING NEEDLE FILE	FIND	5.5	
FN5.5MRK00	5-1/2",00CUT, MARKING NEEDLE FILE	FIND	5.5	
FN5.5B00	5-1/2",00CUT, BARRETTE NEEDLE FILE	FIND	5.5	
FN5.5E00	5-1/2",00CUT, EQUALLING NEEDLE FILE	FIND	5.5	
FN5.5HR00	5-1/2",00CUT, 1/2 ROUND NEEDLE FILE	FIND	5.5	
FN5.5SQ00	5-1/2",00CUT, SQUARE NEEDLE FILE	FIND	5.5	
FN5.5W00	5-1/2",00CUT, WARDING NEEDLE FILE	FIND	5.5	
FN5.53SQ00	5-1/2",00CUT, 3 SQUARE NEEDLE FILE	FIND	5.5	
FN5.5R00	5-1/2",00CUT, ROUND NEEDLE FILE	FIND	5.5	

EXAMPLES: DIE SINKER'S RIFFLERS

Tool Number	Description	Aux.	D/1	D/2
DSR12981-0	MOD.12981,0CUT,6", RIFFLER	RIFF	12981	
DSR12981-2	MOD.12981,2CUT,6", RIFFLER	RIFF	12981	
DSR12965-0	MOD.12965,0CUT,6", RIFFLER	RIFF	12965	
DSR12981-2	MOD.12965,2CUT,6", RIFFLER	RIFF	12965	

TOOL NUMBER CONSTRUCTION GUIDE — *Grinding Wheels*

TOOL NO.: Class Code/Abrasive, Style, Width in Inches, Dia. in Inches, Grade, Grit No.

CLASS	ABRASIVE	STYLE	WIDTH	DIA.	GRADE	GRIT
GW	**A**	**P**	**2W**	**8D**	**L**	**46**
GW	A	P	xW	xD	SOFT	No.
Grinding	Aluminum	Plate	Width in	Diameter	to	Extra
Wheel	Oxide	Mount	x Inches	in x Inches	HARD	Fine
						to
PW	D	C			H	Extra
Polishing	Diamond	Cup			I	Course
Wheel		Wheel			J	600
	B				K	to
BW	Borazon	D			L	12
Burringl	CBN	Dish			M	
Wheel		wheel			N	
	S				P	
CW	Silicon	S			R	
Contact	Carbide	Straight				
Wheel						
	Z	F				
CFW	Zironia	cutoFf				
Cutoff						
Wheel		R				
		Radius				
		U				
		saUcer				

Aux.. I.D.: Whxx D/1: Wheel Diameter D/2: Grit No.

X. Taper Shank Tool Holders and Adapters

Taper shank tool holders are widely used. Most CNC machine tools, except turning and a few others, cannot operate without them. They can also be required for conventional machine tools.

In common use today you will find basically five different types of shanks for taper shank tool holders and adapters: NMTB, Cat V Flange, BT V Flange, R-8, and Quick Change, which can cover a variety of tapers and modifications. Most companies try to standardize their machines to one or two of these holder types, but it is not uncommon to find all of them on the shop floor.

In order to distinguish one holder from another, the ITCS numbering system looks at the following characteristics.

- Shank tapers 30, 35, 40, 45, and 50 are the most common among the different shank styles.
- Shank style or type includes NMTB taper style, Caterpillar V Flange, BT Flange, and R-8 or Bridgeport-type spindles of conventional style.
- There are many different holder types, including collet chuck, end mill, tap, to name a few.
- Tool size means the size of cutting tool or adapter that can be accommodated in the holder. In the case of collet chuck, it would be the max range of the collet style, for example, T G 100 or T G 150 collet. In the case of an end mill holder, it would be the diameter of the bore for the endmill's tool shank.
- Gage length is the distance from the gage line on the shank to the end point or face of the holder.

Constructing Tool Numbers and Aux. I.D.'s for TAPER SHANK TOOL HOLDERS

Considerations Used for General Tool Number Construction

TYPE	STYLE	MODIFIER	MODIFIER
OF			
HOLDER	TAPERS	ADAPTER STYLE	VARIATION
SHANK			
NMTB	30		
CAT V	35	COLLET CHUCK	STANDARD LENGTH
FLANGE		END MILL	EXTENDED LENGTH
	40	FACE MILL	
BT		SHELL MILL	MODULAR
FLANGE	45	SLOTTING	
		MORSE TAPER	EXTENSIONS
	50	JACOBS TAPER	
		TAPPING	SPACERS
R-8	BRIDGEPORT	BORING BAR	
	TYPE SPINDLE	BORING RING	MODIFIED
		BORING HEAD	& SPECIAL ENDS
QUICK	ERICKSON	ARBOR	VENDOR PROPRIETARY
CHANGE		AUTOMOTIVE ADAPTER	

TAPER SHANK HOLDERS & ADAPTERS

Tool No. Begins	General Style of Adapter	TLC Aux. I.D.
xxNMCC	Collet Chuck Adapter	THCC
EM	Endmill Holder	THEM
xxCVFM	Face Mill Adapter	THFM
SM	Shell Mill Adapter	THSM
xxBTSA	Slotting Adapter	THSA
MT	Morse Taper Adapter	THMT
R8JT	Jacobs Taper Adapter	THJT
TA	Tapping Adapter	THTA
xxQCBB	Boring Bar (solid holder)	THBB
BR	Boring Ring Adapter	THBR
xxSPBH	Boring Head Adapter	THBH
AA	Arbor Adapter	THAA
AS	Automotive Shank Adapter	THAS
SP	Special Modified	THSP

Note: The beginning of the tool number specifies the shank taper and type. The two letters that follow this specification indicate the style of adapter for the tool holder.

TOOL NUMBER CONSTRUCTION GUIDE — TAPER SHANK HOLDERS & ADAPTERS

TOOL NO.: ShankTaper/Shank, Style/Adapter, Style/Gage, Length/Tool Size

SHANK TAPER	SHANK STYLE	ADAPTER STYLE	TOOL SIZE	GAGE LENGTH
50	**CV**	**EM**	**.75**	**3.75**
		XXX	Collet Style	X.XX
30	NM - NMTB	CC	T G 100	T G 150
		Collet Chuck	Max. Range of Collet	
35	CV	EM	or	
	Cat V Flange	Endmill	Hole Shank Diameter	
40		FM		S
	BT	Face Mill	Arbor Pilot	Standard
45	BT Flange	SM	Diameter	Length
		Shell Mill		E
50	SP	SA		Extended
	Special	Slotting Adapter		Length
		MT		L
	R8	Morse Taper	Morse Taper No.	Long
	Bridgeport	JT		X
	Type Spindle	Jacobs Taper	Jacobs Taper No.	EXtra
		TA		Long
		Tapping Adapter		Tap Collet No.
		BB		
		Boring Bar	Bore Diameter	
		BR		
		Boring Ring	Pilot Diameter	
		BH		
		Boring Head	Thread Size	
		BA		
		Arbor Adapter	Cutter i.d.	
		AS		
		Automotive Shank	Shank o.d.	
		Adapter		

TAPER SHANK TOOL HOLDERS CLASS OF TOOLS: TOOL DESCRIPTION FORMAT
TOOL SIZE,SHANK STYLE & TAPER,GAGELNGTH,

Aux. Code Format and Dimension Definition

Aux. I.D.: THxx D/1: Shank Taper D/2: Tool Size

EXAMPLES

Tool No.	Description	Aux.	D/1	D/2
40CVCC1003.75	TG100,CAT40VFLNG,3.75GL, COLLET HLDR	THCC	40	100
30NMEM1.002.68	1.00D,NMTB30,2.687GL,ENDMILLHLDR	THEM	30	1.000
45BTFM2.002.12	2.00PILOT,45BTFLNG,2.125GL,FACEMILLAD	THFM	45	2.000
50CVSM0.501.50	.5PILOT,50CATVFLNG,1.50GL,SHELLMILLAD	THSM	50	0.500
40CVSA1.004.00	1.00PILOT,40CATVFLNG,4.0GL,SLOTADRBOR	THSA	40	1.000
35BTMT#32.95	#3MORSETPR,35BTFLNG,2.95GL,MORSETAPER	THMT	35	3
40NMJT#31.50	#3JACOBS,40NMTB,1.50GL,JACOBSCHKADAPT	THJT	40	3
45CVTA1.223.75	1.22DTAPCOLLET,45CATVFLG,3.75GL,TAPHD	THTA	45	1.22
45CVTA#14.56	#1TAPCOLLET,45CATVFLG,4.56GL,TAPADAPT	THTA	45	1
35BTBB2.5011.0	2.50D,35BTFLNG,11.00GL,BORINGBARASSEM	THBB	35	2.5000
40CVBR0.885.38	.875PILOT,CAT40VFLNG,5.375GL,BORERING	THBR	40	0.875
50BTBH87201.75	7/8-20THRD,50BTFLNG,1.75GL,BOREHEADAD	THBH	50	87.20

INDEX